REASON, FAITH, AND PURPOSE

The Ultimate Gamble

REASON, FAITH, AND PURPOSE

The Ultimate Gamble

John R. Fanchi
Texas Christian University, USA

World Scientific
NEW JERSEY • LONDON • SINGAPORE • BEIJING • SHANGHAI • HONG KONG • TAIPEI • CHENNAI • TOKYO

Published by

World Scientific Publishing Europe Ltd.
57 Shelton Street, Covent Garden, London WC2H 9HE
Head office: 5 Toh Tuck Link, Singapore 596224
USA office: 27 Warren Street, Suite 401-402, Hackensack, NJ 07601

Library of Congress Cataloging-in-Publication Data
Names: Fanchi, John R., author.
Title: Reason, faith, and purpose : the ultimate gamble / John R. Fanchi,
Texas Christian University, USA.
Description: New Jersey : World Scientific, [2022] | Includes bibliographical references and index.
Identifiers: LCCN 2021026272 (print) | LCCN 2021026273 (ebook) |
ISBN 9781800610545 (hardcover) | ISBN 9781800610552 (ebook for institutions) |
ISBN 9781800610569 (ebook for individuals)
Subjects: LCSH: Faith and reason. | Cosmology. | Life--Origin. | God--Proof.
Classification: LCC BL51 .F337 2022 (print) | LCC BL51 (ebook) | DDC 210--dc23
LC record available at https://lccn.loc.gov/2021026272
LC ebook record available at https://lccn.loc.gov/2021026273

British Library Cataloguing-in-Publication Data
A catalogue record for this book is available from the British Library.

For any available supplementary material, please visit
https://www.worldscientific.com/worldscibooks/10.1142/Q0311#t=suppl

Desk Editors: Jayanthi Muthuswamy/Michael Beale/Shi Ying Koe

Typeset by Stallion Press
Email: enquiries@stallionpress.com

Printed in Singapore

To Kathy, my partner on this journey

Preface

I have written *Reason, Faith, and Purpose: The Ultimate Gamble* as a guide to eternal questions for believers and inquiring skeptics. This book summarizes the mainstream scientific view of the origins of the universe and life, and analyzes the question of the existence of god from philosophical, religious, and scientific perspectives. If you are a believer, this book will help you understand your faith in a secular world and help you share your faith with nonbelievers. If you are an open-minded skeptic, it will help you answer the question: does god exist?

Reason, Faith, and Purpose: The Ultimate Gamble provides a succinct analysis of key issues and evidence regarding the question of the existence of god. How do we decide if god exists? How did we come to exist? What, or who, created the universe? Is existence simply an accident, or is it part of some divine plan? Answers to questions like these have far-reaching consequences, not just for those who choose to believe in god and choose a religion to fit their beliefs but also for those who decide there is no god, for that decision also influences the way people live their lives.

The material in *Reason, Faith, and Purpose: The Ultimate Gamble* is presented in two parts. Part I presents the secular, scientific view of the origin and evolution of the physical universe and life. It is not enough to discuss the currently popular scientific worldviews because every worldview has blemishes. Consequently, our discussion must include alternative worldviews that help us understand and appreciate differing and, in some cases, conflicting worldviews. Part II introduces other worldviews that are representative of ideas which have been historically prevalent around the world. The material in *Reason, Faith, and Purpose: The Ultimate Gamble* is designed to provide insight into the choice each of us must make in this life: the ultimate gamble.

The literature for the topics discussed in this book is so voluminous that it is necessary to be selective in the choice of material. My goal has been to select material that facilitates the discussion of important, relevant topics in an accessible format for the general reader. Consequently, I have provided references to material that I have found especially noteworthy or that could guide readers to additional information.

Discussion questions are designed to help us develop a better understanding of the relationship between different belief systems and the consequences of choosing a particular belief system. They can be used for self-analysis or by small groups to help readers conduct their own search for truth. *Reason, Faith, and Purpose: The Ultimate Gamble* with associated discussion questions is suitable for a short course or topical study.

Chapter Endnotes — denoted by superscripts in the text — include supplemental comments. References are located alphabetically by author in References and are identified in the text by square brackets as in [author, date]. A section entitled Dates of Selected Characters presents a table of selected people, their countries of origin, and their dates of birth and death. The dates help establish a timeline for the material discussed in the book. I capitalize the word God when referring to the Christian god, call god Allah when referring to the god of Islam, and use the lower-case god to refer to a generic supernatural being.

John R. Fanchi
July 15, 2021

About the Author

John R. Fanchi has a Ph.D. in physics from the University of Houston and is the author of numerous books in the areas of physics, earth science, mathematics, and engineering. He was co-founder and first President of the International Association for Relativistic Dynamics. He has worked in the energy industry and has taught courses in energy, engineering, and physics at Texas Christian University (TCU), Colorado School of Mines (CSM), and The University of Tulsa. His books include *Parametrized Relativistic Quantum Theory* (1993, Kluwer), *Shared Earth Modeling* (2002, Butterworth-Heinemann), *Principles of Applied Reservoir Simulation* (2018, 4th edition, Elsevier), *Math Refresher for Scientists and Engineers* (2006, 3rd edition), *Energy: Technology and Directions for the Future* (2004, Elsevier Academic), *Energy in the 21st Century* (2017, 4th edition, World Scientific), and *The Goldilocks Policy: The Basis for a Grand Energy Bargain* (2019, World Scientific).

Acknowledgments

I would like to thank K. Wayne Fanchi, Anthony Fanchi, Pat Jackson, F. Douglas Shields, and Paul 'Rusty' Lee for insightful discussions over the years. *Reason, Faith, and Purpose* is based on continuing education courses taught at Texas Christian University (TCU), The University of Tulsa, and Arapahoe Community College (Denver area).

Contents

Chapter 1

Quest for Truth

In the beginning, God created heaven and earth.[1]
I am the Lord your God; you shall have no other gods before me.[2]
There is no God but Allah![3]
Religion is the opium of the masses.[4]
God is dead.[5]

Do you find these comments confusing and even contradictory? What should we believe? What is the truth?

Like many others, I have sought answers to questions about the universe and our role in it. I have considered answers based on religious faith and scientific reason. In our search for answers, each of us adopts a worldview or set of beliefs that influences our perspective. Our worldviews can and often do change as we gain more knowledge and experience. I describe my quest here as an illustration of a quest for truth. This background should provide insight into the point-of-view I am bringing to the discussion.

A Personal Quest — Road to Atheism

The worldview I formed as a youth was influenced by my parents' beliefs. My mother was raised a protestant and agreed to convert to Catholicism when she married my father. She tried to raise me as a Roman Catholic in the 1950s and 1960s. As a member of the post-World War II baby boom in the United States, I grew up in a society that was recovering from World War II and learning to live in a world with nuclear weapons. In 1954, Sputnik showed that the two great oceans, the Atlantic and the Pacific, no longer insulated one hemisphere from another. Adversarial nations could deliver nuclear death anywhere in the world in a matter of hours. I still remember learning to "duck and cover" under my desk in my elementary school classroom. Two of the nuclear nations, the Soviet Union and China, were governed by political organizations that rejected religion and God.

Social norms were also changing. We were living through the Civil Rights movement, the sexual revolution, and the widespread use of recreational drugs. Mass media, particularly television, was becoming part of common culture and could be found in almost every home. Competing worldviews were no longer a curiosity saved for academicians and politicians, but were directly influencing all segments of society through the increasingly affordable magic of mass media.

I first began to question my faith when I was in Adams City Junior High School in the 1960s. The school was in Commerce City, a suburb of Denver, Colorado. Commerce City is now home to the Colorado Rapids of Major League Soccer. In the 1960s, Commerce City was the poor side of town with a couple of small refineries, a grain elevator, and an A&W drive-in restaurant.

My youthful questions about faith were met with indifference by priests at my church. One priest suggested that I was possessed by evil for even asking such questions. I rejected this view and decided to look outside of organized religion for answers. I turned to science.

I began by reading books on chemistry and astronomy in 9th grade. I was allowed to use a chemistry lab after school to learn more about chemistry. One of my ideas was to generate carbon dioxide by heating gasoline. Fortunately, my chemistry teacher stopped me before I blew up the lab. In another one of my experiments, sulfuric acid was sprayed out of a flask. This time I went home with a few acid holes in my winter coat. They were small, though, and I was able to keep wearing the coat. I began to develop an appreciation for the hazards of confronting the unknown.

The astronomy book I chose to read, *Frontiers of Astronomy* [Hoyle, 1955], was written by British astronomer Fred Hoyle (1915–2001). Hoyle would eventually be considered a renegade scientist because he chose to reject the conventional views of mainstream scientists. At the time, I chose Hoyle's book because the inexpensive paperback was written by a Professor of Astronomy at Cambridge University, had many pictures in it, and fit my budget. Hoyle spent much of his professional life arguing that the Big Bang theory of cosmology was incorrect and supported a steady state theory of cosmology. In fact, Hoyle coined the term "Big Bang" in a BBC radio broadcast in 1949 in an attempt to contrast the Big Bang theory with his steady state theory. Today, the Big Bang theory is a key component of the scientific worldview.

After my junior high experience, I entered Adams City High School and took college prep courses in biology, chemistry, and physics, in that order. In a way, the order made sense because we were presented a reductionist view of science from the scale of life to the scale of atoms and molecules to the scale of particles. Between this set of courses and my extracurricular reading, I came away with the view that biology could be thought of as an application of chemistry, and chemistry could be thought of as an application of physics. The fundamental science was physics, and it was only a matter of time before science understood everything there was to know about nature.

I was impressed by what scientists had learned about nature. After all, it was widely believed that physicists played a major role in ending World War II by using their knowledge to develop an atomic bomb. Physicists not only understood how the Sun worked but also could use their knowledge to develop the hydrogen bomb, which was more powerful than the two atomic bombs that were dropped on Japan in 1945. Even science fiction was coming of age with the broadcast of the first Star Trek series in the late 1960s. Episodes of Star Trek explored the relationship between logic and emotion, and cast scientists as both explorers of the final frontier and as warriors who fought only when it was necessary to protect the common good. By the time I graduated from high school in 1970, I was an atheist.

A Personal Quest — The Winding Road

I attended the University of Denver on a full-tuition scholarship from 1970 to 1974. As a physics major, I was required to take two minors to graduate with a B.S. degree in physics. I chose to minor in math and chemistry. These minors were typical choices for undergraduate physics majors. I was informed by my advisor after my first quarter, though, that I had tested out of chemistry. I could choose another minor, so I decided to minor in psychology. I soon learned the difference between hard and soft science.

My physics advisor was not happy that I chose psychology as a minor because psychology was considered a soft science. It was more of a descriptive science than a quantitative discipline like physics and, to a lesser extent, chemistry. A special set of psychology courses had to be selected that the faculty would consider appropriate for a B.S. degree in physics. Each psychology course in my minor had to have either a mathematical component, like statistics, or a laboratory component. The psychology minor gave me a chance to participate in experiments with living organisms. In my case, I worked with rats and snails and participated in a study about hypnosis.

Why did I choose to minor in psychology? I think some faculty members thought that I chose psychology because there were almost no women in physics classes, and there were many women in psychology classes. In reality, though, as a result of my extracurricular reading in relativity and quantum mechanics, I had learned that fundamental physics was concerned about the observer. In relativity, the form of the equations that describe a physical system can depend on the motion of the observer. In quantum mechanics, the observer can alter the state of the system by the act of observation. The observer, such as you or me, was an important part of the system. I thought I needed to learn something about the observer, and psychology is the study of human behavior, including emotion, sense, and perception. It seemed like a logical choice at the time and proved to be helpful 20 years later when I served as Vice President of the School Board in Littleton, Colorado.

During the four years I spent at the University of Denver, I worked halftime at the Denver Research Institute conducting experiments with an ultrasonic interferometer. The money, though small, supplemented my scholarship and helped pay my living expenses. The ultrasonic interferometer was an instrument that measured the speed of sound in gases. These measurements were used to test quantum mechanical theories that related the speed of sound to vibrations and rotations of gas molecules. This research in chemical physics led to my first peer-reviewed publication and helped me get a graduate assistantship at the University of Mississippi — also known as Ole Miss — where related work was being conducted.

Personal Anecdote: The Ultrasonic Interferometer

I had a memorable experience with the interferometer a couple of years after I began using it to make measurements. The interferometer included a metal vessel that held a liquid that could be heated to create a heat bath. The liquid was circulated by a small mixer to keep the temperature in the bath relatively constant. One day I came to work and discovered that the liquid in the bath had become a highly viscous gel that could only be scraped out by hand. This was, of course, my task. This experience helped me appreciate why experimentalists calibrate their apparatus before using the apparatus to make new measurements. Calibration lets them verify that the apparatus is working as designed and that nothing has gone wrong, such as the unexpected formation of a gel. The calibration step helps assure experimentalists that their results are reproducible. I learned an important principle: a physical system can work as expected for a long time, but it requires an act of faith to assume the system will continue to function in the future the same way it has functioned in the past. My experience involved an experimental apparatus, but the principle also applies to scientific theories.

The undergraduate physics curriculum at the University of Denver consisted primarily of courses that just touched on the more interesting topics in physics from my point of view. I pursued my true interest in physics with outside reading and independent studies with a couple of willing physics faculty members. Physics relies heavily on mathematics, and my mathematical skills had advanced to the point where I could read some of the more interesting books on relativity and quantum mechanics. One of the books was *The Mathematical Theory of Relativity* [Eddington, 1923], written by English astrophysicist Arthur Eddington (1882–1944). He was one of the few people who first understood the general theory of relativity [Einstein, 1916] published by German physicist Albert Einstein (1879–1955). Eddington helped make Einstein famous by observing a total solar eclipse in 1919, shortly after World War I, and verifying Einstein's 1911 prediction that the trajectory of light would be curved by the presence of a massive object such as the Sun. Like Hoyle, Eddington became a renegade scientist later in his life when he tried to unify the various theories of physics.

Unlike Hoyle and Eddington, Einstein began his career as a renegade physicist. Einstein could not find a teaching post in physics when he graduated. He had to work outside of the physics mainstream like a role model of his, seventeenth century Dutch philosopher Baruch Spinoza (1632–1677). Spinoza believed that god was the only substance in nature. He made no distinction between the creator and its creations. To Spinoza, all things were merely different forms of one substance. If everything in nature was nothing but a different manifestation of God, Spinoza argued, then the more traditional explanations, including those advocated by contemporary Jewish and Christian authorities, were unnecessary. Although Spinoza believed in a god, it was not the God of his Jewish community. Spinoza was censured by the Jewish community (similar to ex-communication) and forced to earn his living as a lens grinder rather than a philosopher. Lens grinding was considered on the forefront of technology during Spinoza's time. Spinoza demonstrated that a scholar could earn a living outside of the intellectual establishment and still contribute to humanity's store of knowledge. This was an important lesson for Einstein,[6] who also agreed with Spinoza's metaphysics. Einstein said in a telegram sent to a Jewish newspaper in 1929 that "I believe in Spinoza's God who reveals himself in the harmony of all that exists, but not in a God who concerns himself with the fate and action of men" [Calder, 1982, p. 138].

Some of Einstein's most influential work in theoretical physics was conducted in his free time while he was a clerk in the patent office in Bern, Switzerland. Einstein was not noticed by the physics community until after 1905, the year he published an explanation of the photoelectric effect, an explanation of Brownian motion, the special theory of relativity, and a related paper that showed the equivalence between matter and energy. The work initially drew little attention or was rejected by mainstream physicists, typically physicists who occupied influential academic positions. Eventually, the physics community accepted some of Einstein's ideas. Einstein became a full-time academic when he accepted an Associate Professor position at the University of Zurich in 1911. Rather than following the mainstream worldview, Einstein pioneered a new path for the mainstream to follow.

When I graduated with a B.S. in physics in 1974, I had minors in math and psychology and a sense that something was amiss. After four years of a liberal arts and science education, I was beginning to realize that human knowledge was incomplete. Quantum mechanics had replaced Isaac Newton's (1643–1727) classical mechanics with a theory that relied on Werner Heisenberg's (1901–1976) uncertainty principle and Max Born's (1882–1970) probability interpretation. Prior to the development of quantum mechanics, many philosophers had argued that Newton's mechanics implied a deterministic universe. The only role God had in such a clockwork universe was to establish the initial conditions that set the universe in motion. In a deterministic universe, every action is decided once initial conditions are specified. Did quantum mechanics challenge the notion of a deterministic universe and provide scientific support for the notion of free will? Do we know with certainty how the universe began and how life evolved? By the time I arrived at Ole Miss, I had to admit that I did not have all the answers. I had become an agnostic.

I spent a year and a half at Ole Miss. In addition to taking classes for my M.S. in physics, I taught undergraduate labs and conducted research studying the effects of laser light on chemical reaction rates. Since there was a lack of space in the main physics building on campus, my lab was on the ground floor of the old observatory behind the physics building. The laser I used for my research had to be built from scratch using materials that were available in the department. My advisor did most of the design and construction, but I had a chance to polish metal mirrors until they were optically flat. These mirrors were used to confine the laser light. A small hole drilled in one mirror allowed the light to escape and form a beam that could be directed at our reaction chamber. When the laser was working, the frequency of the laser light was designed to excite a particular vibrational mode of a molecule. This experiment was an application of quantum mechanics in the real world at a time when the United States and the Soviet Union were in the middle of the Cold War. The use of lasers to alter chemical reaction rates interested the U.S. government in the 1970s as a possible method of enriching uranium for nuclear weapons. My work led to co-authorship with my advisor of a peer-reviewed paper.

Personal Anecdote: Experimental Safety

Experimental research in academia is supposed to follow government guidelines, but when you are working on the frontier of knowledge, it is not always easy to find applicable guidelines. The homemade laser we were using in the mid-1970s had exposed metal surfaces that were electrified when the laser was in operation. I had safely performed many experiments with the laser-chemical reactor system. One day, however, I found myself extending a finger toward one of the electrified metal surfaces. I could see my finger inching closer to the surface even though I was thinking that I should stop. One moment I was standing next to the table supporting the laser and reaching toward the metal surface, the next moment I found myself a foot away from where I was originally standing. The shock had apparently made me jump. Had I been grounded, the shock could have been fatal. Was I lucky in an accidental universe, or was I spared in a purposeful universe?

I had two peer-reviewed papers in chemical physics by the time I was 22. In both cases, I saw aspects of academic science that showed me the human side of the enterprise. Publication of the paper using my work at the University of Denver was delayed a few months as a penalty for us not paying page charges. Some science journals, including the most respected journals, charge authors the cost of publishing their work once the papers have been accepted by anonymous referees. Federal grants for research may authorize grant money to pay these charges. If there had not been a delay, my first paper would have appeared when I was still a teenager.

My second paper reported my Ole Miss work. It was published in the mainstream Journal of Chemical Physics, but I did not write a thesis using the work. I left Ole Miss with an M.S. in physics based on coursework. I could have graduated with an M.S. in physics with a thesis if I was willing to pay the cost of tuition for thesis credit. I was learning that money was an important part of academia and science.

The next stop in my quest was the University of Houston. I was awarded a teaching fellowship to continue graduate work toward a Ph.D. in physics.

Rather than continuing with experimental research, I initiated a theoretical research program into the foundations of physics that I have continued for over 40 years. I made a living teaching laboratory classes while I attempted to construct a new foundation for relativistic quantum theory. I wrote in the Preface to my Ph.D. dissertation [Fanchi, 1977], "My principal motivation for undertaking this work was the desire to answer the question: can free will and determinism coexist? The answer to this question has important consequences in fields other than physics, e.g. theology." By mixing nonconforming science and philosophy, my research was placing me firmly in the camp of renegade physicists. This gave me a special vantage point to continue my quest for truth.

Personal Anecdote: International Association for Relativistic Dynamics

I have continued my physics research program since 1977 without being dependent on a physics establishment that is now dominated by a secular perspective. Indeed, by 1998, I was able to convene and co-found a new organization in physics that is known today as the International Association for Relativistic Dynamics (IARD). The association had its first meeting in Houston and has met every other year since 1998. The purpose of IARD is to facilitate "the acquisition and dissemination of knowledge about research programs in classical and quantum relativistic dynamics of particles and fields."[7]

I had an important choice to make when I graduated with my Ph.D. in physics in 1977. I was offered a position at the U.S. Naval Research Laboratory (NRL), a commission into the U.S. Nuclear Navy, or a postdoctoral position studying the geothermal storage of solar energy. The NRL position was offered to me because of the laser chemistry work I did at Ole Miss. It was part of the effort to win the Cold War. The U.S. Navy commission was a choice between a short-term teaching position in Orlando, Florida, or a career at sea. The position that gave me the best chance of continuing my quest was the postdoctoral position, which led me into the energy industry a year later.

Working in the corporate world has advantages and disadvantages. Some of your time is not your own. Of course, academia has this problem as well. When I was a professor in academia, I was evaluated on three components: teaching, research, and service. The relative importance of each component depends on the university. In many cases, less than half of an academic evaluation depends on teaching. Academics argue that research is what makes a university a community of learning, and to an extent, research is a teaching activity because the faculty member is expected to conduct research with students. The service component includes committee work for the academic department, the university, and external professional organizations. Industry may have demanded more of my time during the workday, but I usually had more free time during evenings and weekends.

It is common in industry to give up intellectual property rights for work-related activities. The loss of intellectual property rights is becoming an issue for academics because university and government administrators now realize the potential value of these rights to their general funds, but there is usually more freedom of inquiry in academia than in industry. The trick is to work for an industry where there is no conflict of interest.

Even though academics may enjoy more freedom of inquiry in principle, many universities expect their faculty to obtain external funding. For a good evaluation, which influences pay raises, promotions, and access to resources and other administrative support, faculty members must find someone to fund their work or else spend their time doing work someone is willing to fund. Faculty evaluation procedures in the modern university have given people who control funding sources in government agencies and private industry considerable power over faculty research, often at the expense of academic freedom. If you have a position in industry that pays better than academia, it is possible to use a corporate salary to fund your own work, especially if you are conducting theoretical research. Financial independence makes it possible to pursue research that would otherwise be considered out of favor or nonconforming, but has the potential for revolutionizing our understanding of nature. I was able to find the time and resources to continue my quest while working in industry.

Using my home address as my institutional byline, I wrote and published a peer-reviewed article in physics every year or two. Most full-time academics would try to publish at a higher rate, but I was not obligated to function in a "publish or perish" environment. Besides, publications in other subjects and peer-reviewed journals let me compete in the academic arena. Each of my publications in theoretical physics was designed to advance my research program. By 1993, I had published enough peer-reviewed papers to write my first book, which was entitled *Parametrized Relativistic Quantum Theory* [Fanchi, 1993]. A few years later, in 2001, my research in physics and my multidisciplinary background were factors that helped me earn tenure as a professor at the Colorado School of Mines and later at Texas Christian University.

Personal Anecdote: Tenure Criteria

I have had the opportunity to navigate the tenure process as well as serve on faculty committees that evaluated tenure and promotion applications. Universities may grant different forms of tenure, and the criteria can vary from one university to another. Typically, a tenured Professor is expected to have a Ph.D. in the appropriate academic discipline, an international reputation, and a viable research program. Reputation can be measured by letters of recommendation from ranking faculty at other institutions, preferably with an international representation. The viability of a research program can be measured by how much money you raise and the number of publications you generate. These criteria are worth noting because they demonstrate that leaders of a field can influence the careers of other academicians, even at institutions in other countries. If you are seeking tenure and promotion, it can be harmful to your career to be considered a renegade.

I was broadening my knowledge by necessity and by choice. To succeed in the industry, I had to learn about the real-world application of physics in engineering and geosciences, especially geology and geophysics. I found that there was often more than one model that could fit natural data. I documented some of these concepts in *Shared Earth Modeling* [Fanchi, 2002] and *Principles of Applied Reservoir Simulation*, fourth edition [*Ibid.*, 2018], and taught undergraduate and multidisciplinary graduate courses using these textbooks.

My outside reading included philosophy and theology, especially books on comparative religion and creation stories. To test my understanding and my knowledge, I began to write drafts of this book and teach adult education classes about the relationship between science and religion. People attending my classes over the years have ranged in age from their 20s to retirement age, and they came from many walks of life.

The first adult education class I taught on science and religion was through the Praxis Project at The University of Tulsa in the early 1980s. A couple attending the class was thinking about marriage but did not share common religious beliefs. The woman was a believer, and the man thought he was an atheist, but later realized he was an agnostic. By the time the class was over, they had a much better understanding of their beliefs and the life they could share together. Although I had already learned much about the abstract nature of spirituality, this incident helped me appreciate the importance of spiritual issues in some people's lives.

Personal Anecdote: Full Disclosure

I mentioned earlier in this chapter that I was raised a Catholic, chose to become an atheist, then realized that I had enough questions left to become an agnostic. My life experiences have led me back to theism and I now consider myself a Christian.

My goal is to present as objective an appraisal as possible of the material you will need to consider the question of god's existence. This presentation will include discussion of questions that remain unanswered by science and identify the shortcomings of various worldviews. The discussion will be influenced by my perspective, which is why I am disclosing my personal worldview. I hope this disclosure will help you better appreciate the discussion of the questions we are considering.

Religious Beliefs

My journey has shown that worldviews can transform from one belief system to another. Different religious belief systems are described in Table 1.1.

Table 1.1 **Religious belief systems.**

Theism	Belief that a supernatural intelligence created the universe and is involved in human affairs (e.g. a personal god)
Deism	Belief that a supernatural intelligence established the laws that govern the universe
Pantheism	Belief that god is equivalent to nature
Agnosticism	Belief that we do not know if god exists
Atheism	Belief that there is no god

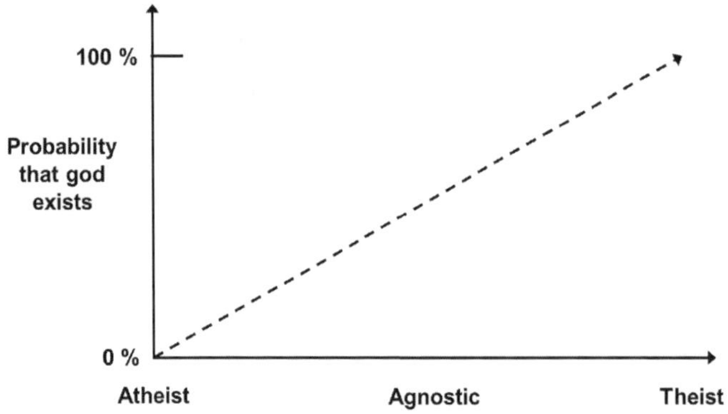

Figure 1.1. Sketch of belief system versus perceived probability that god exists.

Figure 1.1 is a sketch of the relationship between three belief systems and the probability that god exists. The dashed line is intended to show that we do not know if the relationship is linear or nonlinear. Most people tend to have some uncertainty about their beliefs so we use a scale that ranges from 0% probability that they believe in the existence of god to 100% probability. By god, we are referring to a single god, godhead, or multiple gods. A pure atheist would say that there is a 0% probability that god exists while a pure theist would say that the probability that god exists is 100%. Most people believe that god exists with a probability that is somewhere in between these two extremes. Where are you on the curve?

We are focusing here on the relationship between science and religion. From this point of view, it is fair to ask, do scientists believe in a personal god? Surveys have been conducted to answer this question. Table 1.2 is a compilation of survey answers among "greater" scientists, where the term

Table 1.2 Do you believe in a personal god?

Survey Year	1914	1933	1998
Personal belief (percent)	27.7	15	7.0
Personal disbelief (percent)	52.7	68	72.2
Doubt or agnosticism (percent)	20.9	17	20.8

"greater" refers to members of the U.S. National Academy of Sciences (NAS). The authors Larson and Whitman [1998] showed results for surveys conducted in 1914, 1933, and 1998. They concluded that their survey "found near-universal rejection of the transcendent by NAS natural scientists." In addition, the trend from 1914 to 1998 showed that belief in a personal god has been declining. If you combine the percentages for personal disbelief and doubt or agnosticism, approximately 93% of survey respondents doubt or disbelieve that there is a personal god. It is worth noting that members of NAS are often involved in setting national education standards that are subsequently imposed in school districts throughout the United States. For example, Larson and Witham pointed out that NAS issued a booklet [NAS, 1998] that encouraged the teaching of evolution in public schools and assured readers that science is neutral about the question of the existence of god.

Larson and Witham also asked if NAS members believed in immortality. Their results are shown in Table 1.3. The trend from 1914 to 1998 showed that belief in human immortality by NAS members has been declining.

Ecklund and colleagues [Ecklund *et al.*, 2016] asked a broader range of scientists about their religious views. They asked biologists and physicists about their perceptions of the science-faith interface. Over 9400 scientists responded to the survey. The respondents were at all career stages in eight regions of the world: France, Hong Kong, India, Italy, Taiwan, Turkey, the United Kingdom, and the United States. The scientists represented elite and nonelite universities and research institutes. Ecklund *et al.* found that scientists tend to have more secular views than their general populations. In four of the regions — India, Italy, Taiwan, and Turkey — the majority of scientists considered themselves religious. Most of the scientists in the survey thought that science and religion operated in separate spheres. The view that science and religion are in conflict is more pronounced in Western regions.

Table 1.3 Do you believe in human immortality?

Survey Year	1914	1933	1998
Personal belief (percent)	35.2	18	7.9
Personal disbelief (percent)	25.4	53	76.7
Doubt or agnosticism (percent)	43.7	29	23.3

Geologist Stephen Jay Gould (1941–2002) suggested that science and religion were not in conflict, but were actually interested in different domains of teaching authority that do not overlap.[8] To Gould, the scientific domain was focused on facts, while the religious domain was focused on values. Gould's view was known as NOMA, or nonoverlapping magisteria, where a magisterium was a teaching domain.

The New Atheists

Modern secular scientists are beginning to explicitly express their views as atheistic views. A book that illustrates the thinking of the new atheists is *A Universe from Nothing* [Krauss, 2012]. The author Lawrence Krauss points out in the beginning of his book that it is a layman's candid guide to the new atheism being espoused by a segment of the scientific community. Richard Dawkins, a British academic and atheist, wrote a preface for Krauss' book. Dawkins likened Krauss' book to Charles Darwin's (1809–1882) work on evolution. Darwin's work made it acceptable to question whether God created life, while Krauss is making it acceptable to question the need for a supernatural creator of the physical universe. We consider some of these ideas in later chapters to develop a better understanding of the new atheism.

An earlier expression of a secular view was expressed by physicist Steven Weinberg, who wrote in the first edition of his book *The First Three Minutes* that "The more the universe seems comprehensible, the more it seems pointless" [Weinberg, 1977, p. 144]. It is worth noting that Weinberg published this comment in 1977, and then complained that the "phrase has dogged me ever since" [*Ibid.*, 1992, p. 255]. He tried to clarify his comment when he explained that he "did not mean that science teaches us that the universe is pointless, but rather that the universe itself suggests no point" [*Ibid.*, 1992, p. 255]. Weinberg had the chance to edit the comment when he updated his 1977 book in 1993, but he chose to keep the comment unchanged despite the negative feedback [*Ibid.*, 1993, p. 154]. According to Witham [2003, p. 153], Weinberg said that naturalism was better than religion because religion was tantamount to a belief in fairies.

The new atheists have been challenged by their peers. For example, Gerald Schroeder (1938–) related Biblical and scientific chronologies of the origin of the universe and life in *The Science of God* [Schroeder, 1997, p. 67]. More recently, Amir Aczel (1950–2015) rebutted many of the arguments being expressed by new atheists [Aczel, 2014].

An important point to make about the debate between theists and atheists is that the existence of god does not depend on whether or not we can win a debate. One side of the debate might be able to persuade their audience to agree with them, but that does not change the reality of the existence of god. Either god exists, or there is no god.

Figure 1.2 illustrates some key issues associated with the origin of the physical universe or life. Beginning at the top of the figure, we first hypothesize that the universe either had no beginning or else it had a beginning. An eternal universe would not require a beginning. If the universe had a beginning, we can ask whether it began with cause or without cause. A universe could begin without cause if the universe appeared as an accident of nature. An accidental creation can be thought of in scientific terms using

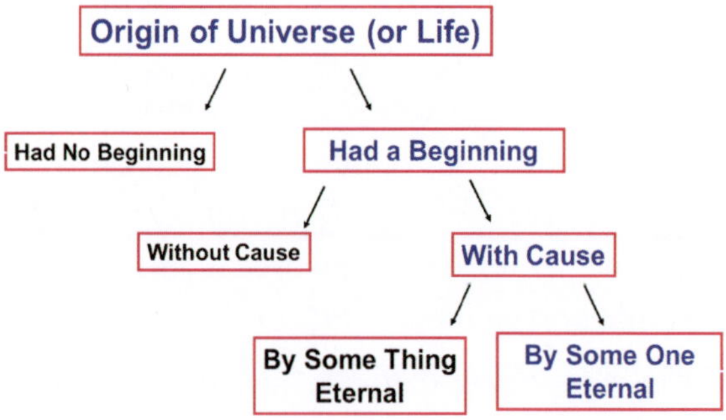

Figure 1.2. Key issues.

such physical concepts as a quantum foam and multiverse. These concepts are discussed in more detail later. If the universe began with cause, we then ask if the cause was by something eternal, or by someone eternal. This gets us into the realm of religion and the supernatural.

The above discussion could be repeated by replacing the word "universe" with "life," as suggested in the figure. Our focus here is to address these issues from a scientific perspective. Later we look at the issues from more philosophical and religious perspectives.

Scientists believe that the physical universe began with a big bang, and new atheists believe that the big bang occurred without a supernatural creator. Is the Big Bang fact, theory, or narrative? The Big Bang is a scientific theory that is based on some facts and is the mainstream paradigm, or narrative, of modern cosmology.

Thomas Kuhn (1922–1996) introduced the concept of paradigm as part of his theory of scientific revolutions [Kuhn, 1962]. Kuhn's work helped motivate the modern study of scientific revolutions. Other theories of scientific revolutions are discussed in Chapter 13. Kuhn argued that science is characterized by belief systems he called "paradigms." A paradigm is a set of beliefs regarding a particular aspect of nature. Paradigms have several functions: they determine the structure of knowledge; they help scientists decide which problems are most important; and they provide a conceptual foundation upon which explanations of natural phenomena are based. The latter two functions play an important role in the day-to-day activities of what Kuhn calls "normal science."

One goal of normal science is to test predictions made within the context of the paradigm. Another goal of normal science is to extend the applicability of the paradigm to new areas. Efforts to reformulate the paradigm into a more aesthetic form are also acceptable activities of normal science. The paradigm of normal science is the standard against which competing paradigms are measured.

Normal science is the mainstream of scientific activity: its practitioners receive the greatest amount of research funding; they control educational institutions; their paradigm is the one taught in most textbooks; and they control access to journals that are considered reputable by other mainstream scientists. University faculty and journal editors are often the gatekeepers of science. They decide what is suitable and what is not.

Personal Anecdote: A New Academic Program

I had a chance to experience first-hand the effort required to enact change in academia when I was a Professor at the Colorado School of Mines. I developed a new course and accompanying textbook that introduced science and engineering students to the emerging energy industry [Fanchi, 2004]. The course helped launch a new multidisciplinary minor in Energy for undergraduate students. Approval for the minor required securing approvals from the host academic department, the Undergraduate Council (a university-wide committee that is responsible for the undergraduate curriculum), the Faculty Senate (a university-wide representation of faculty), the administration, and the Board of Trustees. The concern expressed most often was how the minor would impact existing academic majors. The approval process was arduous because it helped preserve the quality of the academic program, but it had the secondary effect of hindering innovation.

Work outside of normal science is often considered fringe science [Friedlander, 1995]. People who decide to work outside of normal science are nonconforming and may be viewed as renegades. Fred Hoyle, Arthur Eddington, and Albert Einstein are examples of people who were renegades at some point in their careers.

What Lies Ahead?

Shared worldviews can bind human beings together, while differences in worldviews can lead to conflict. Today we are seeing worldviews clash on local, national, and global scales. Examples include the secularist challenge to Judeo-Christian traditions in the United States and Europe, and Jihadists seeking to replace the influence of Western democracies with Islamic rule. These clashes are playing themselves out in a variety of venues ranging from courtrooms to battlefields. Many clashes occur because people cannot agree on an answer to the question: does god exist?

In this chapter, I outlined my motivation for writing this book and presented some events that shaped my perspective. In addition, I discussed the range of religious beliefs in today's world and introduced Kuhn's theory of scientific revolutions.

The scientific view of cosmology and life is presented as a secular worldview in Part 1. Chapters 2 through 5 provide information that is needed to understand the scientific view of the Big Bang and its aftermath. Chapters 2 and 3 present evidence for the expanding universe and inflationary cosmology respectively. Chapter 4 introduces the Standard Model of particle physics and presents a mechanism for initiating the Big Bang. Chapter 5 describes the evolution of the physical universe. The origin and evolution of life are described in Chapters 6 through 8.

Part 2 examines the scientific worldview from the point of view of religious and philosophical perspectives. Chapters 9 through 12 review mythological and religious perspectives, and include synopses of the world's major religions. Metaphysical and epistemological perspectives are discussed in Chapters 13 and 14 with a focus on cosmology and the veracity of the scientific method.

The origin and evolution of the universe and life are sources of conflict between competing worldviews. Chapters 15 and 16 summarize the competing worldviews, and their merits are reviewed in the context of theistic, atheistic, and humanistic perspectives. The choice of a worldview and associated beliefs is discussed in Chapter 17. We are reminded that our choice of worldview can have eternal consequences.

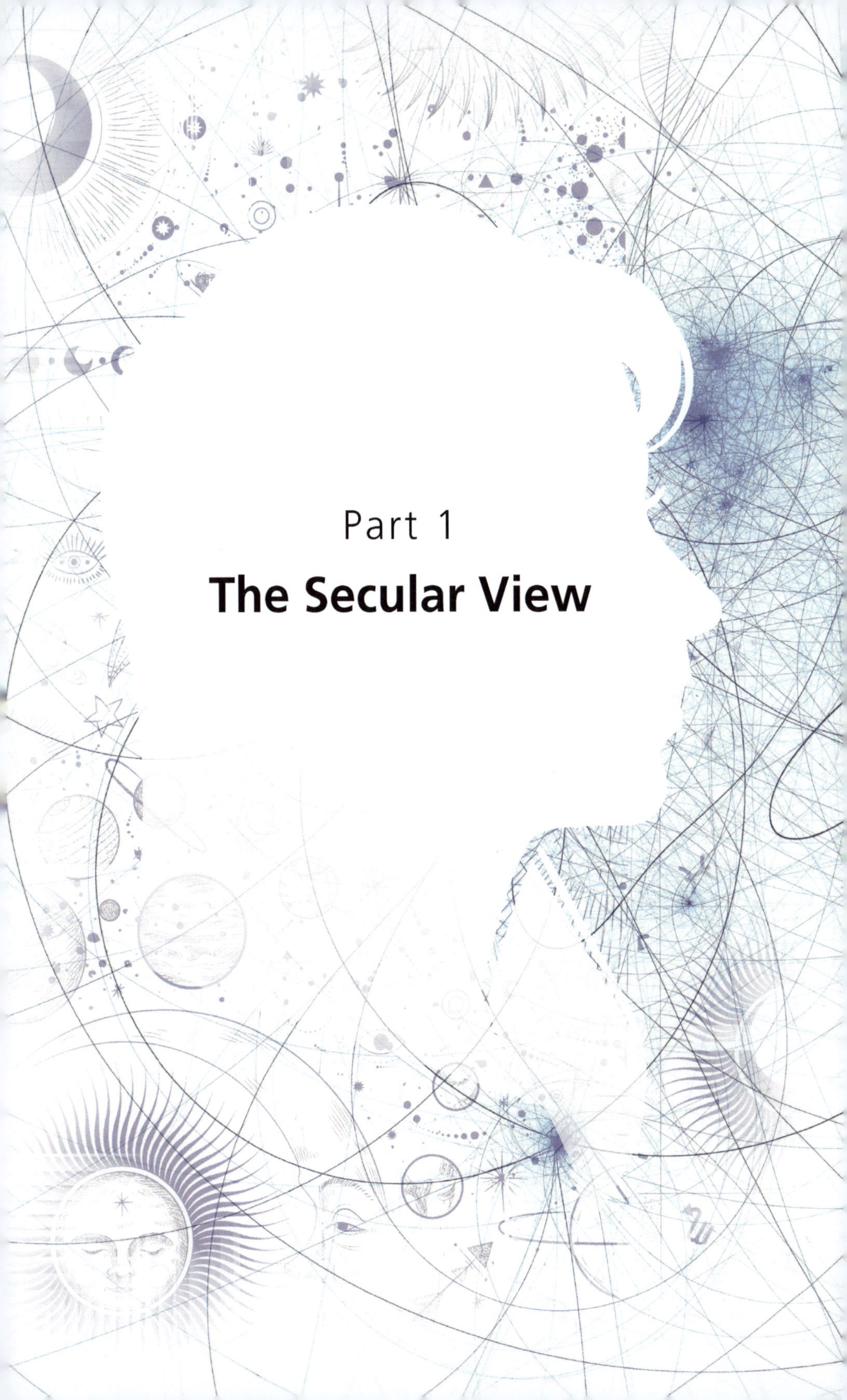

Part 1
The Secular View

Chapter 2

Evidence for an Expanding Universe

Modern cosmology is based on the idea that the universe began with an explosive release of matter and energy known as the Big Bang. Scientists believe that we live in an expanding universe that exists in the aftermath of the Big Bang. In this chapter, we examine the evidence for an expanding universe beginning with the discovery of galaxies.

The Hubble Diagram

The realization that the universe contained more than one galaxy occurred early in the twentieth century. Astronomer Edwin P. Hubble (1889–1953) used a 100-inch telescope at the Mount Wilson Observatory in California to observe the motion of astronomical objects in the sky during the 1920s and 1930s. Hubble prepared a diagram showing the radial velocity of a galaxy in km/sec plotted against distance from Earth to that galaxy in megaparsecs. A megaparsec is approximately 3.26 million light-years, and a light-year is the distance light travels in a year. The diagram is now called the Hubble diagram and examples are shown in Figure 2.1.

Figure 2.1(a) shows Hubble's velocity-distance relation using data from his 1929 paper [Hubble, 1929]. Figure 2.1(b) shows Hubble and Humason's velocity–distance relation using data from their 1931 paper [Hubble and Humason, 1931]. The dots at the lower left are data points from Hubble's 1929 paper. Figure 2.1(b) includes the velocity of astronomical objects at much greater distances than the observations presented by Hubble in his 1929 paper. Hubble and Humason [1931] used the redshift to observe more distant objects. The redshift is discussed below.

Hubble observed a linear correlation that shows that galaxies farther from Earth are moving away from us at a greater velocity than galaxies that

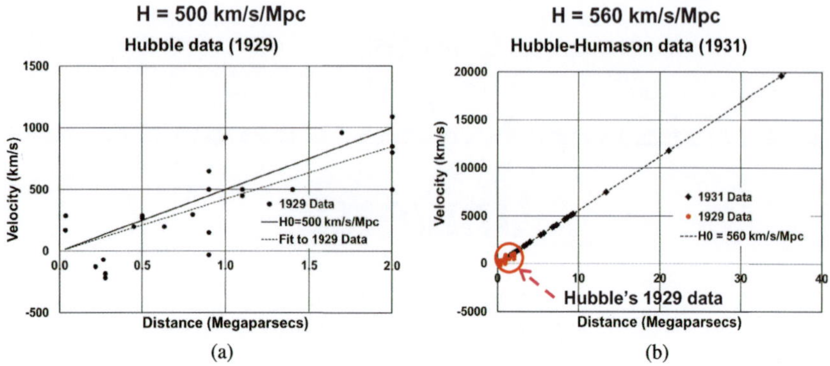

Figure 2.1. Hubble diagrams.

are closer to Earth. For example, the dashed line in Figure 2.1(a) is a linear regression fit to the data, while the solid line is based on the estimated slope H_0 of 500 km/s/Mpc where 1 Mpc is a megaparsec. By contrast, the dotted line in Figure 2.1(b) is based on the estimated slope H_0 of 560 km/s/Mpc. The slope H_0 of the line drawn through the points is now known as the Hubble constant, or Hubble's parameter at the present time.

Hubble realized that the Milky Way is not the only galaxy in the sky; that some "nebulae," or gas clouds, were actually galaxies containing stars; and that the galaxies are moving away from each other. The slope H_0 gives the rate of expansion of the universe. Hubble and Humason [1931] estimated the rate of expansion of the universe to be 560 km/s per megaparsec. The modern estimate of the rate of expansion is much lower, as we explain in the following.

Point to Ponder: Observing the Past

We are looking into the past when we observe light from astronomical objects such as stars or galaxies. The speed of light is 300,000 km/s or about 186,000 miles per second. This may seem like a large speed, but it is finite. It takes time for light to reach us. For example, Alpha Centauri is the closest star system to our Solar System. It is a system of three stars, and is about 4.4 light-years away. A light-year is the distance light travels in a year (approximately 9.5 trillion km, or almost 6 trillion miles). Light from a distant star will take longer to reach us than light from a nearby star.

Cosmological Redshift

How do we know that galaxies are moving away from each other? One of the tools that scientists use is the cosmological redshift, which is an optical version of the Doppler effect for sound.

Sound is the wave-like transmission of vibrational energy. Since it is wavelike, sound has a wavelength and frequency (or pitch). If the wavelength of sound changes with respect to an observer, so also will its pitch. As an illustration, imagine standing next to a railroad crossing as a train approaches the crossing (Figure 2.2).

When the train gets close to the crossing, the engineer blasts the train horn in warning. The sound of the horn begins before the train enters the crossing and continues until the train engine exits the crossing. To the engineer, the sound of the horn never changes because the engineer is moving with the train. To those of us standing next to the crossing, the sound of the horn rises in pitch as the train approaches, and then falls as the train recedes. This change in pitch was first explained by Austrian physicist and mathematician Christian Andreas Doppler (1803–1853).

The train engineer does not hear a change in pitch because the sound of the horn and the engineer are stationary relative to each other. The change

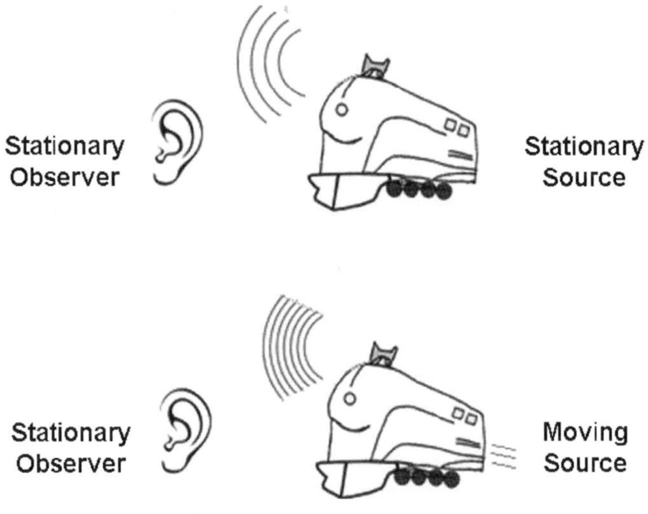

Figure 2.2. Doppler effect for sound.

in pitch we hear is the Doppler effect for sound. As the train approaches the crossing, the number of sound waves reaching us in a given time interval increases. This is the cause of the increase in pitch we hear. When the train passes and recedes, the number of waves reaching us in a given time interval decreases so that we hear a decreasing pitch. Changes in wavelength or pitch can result from the relative motion of the source of wave motion and an observer.

The Doppler effect applies to any wave phenomenon. It states that the frequency, or equivalently, the wavelength, of a moving wave depends on the motion of the observer. Suppose we consider a source of light, such as a star, that is being observed by a stationary observer. If the star is not moving relative to the observer, the wavelength of the emitted light is illustrated in Figure 2.3(a). If the source of a wave of light is receding from a stationary observer, the wavelength of waves passing the observer will increase, as shown in Figure 2.3(b). The increase in wavelength associated with a receding light source is known as a redshift. By contrast, if the source of a wave of light is approaching a stationary observer, the wavelength of waves passing the observer will decrease. In the case of an approaching light source, the decrease in wavelength is known as a blueshift.

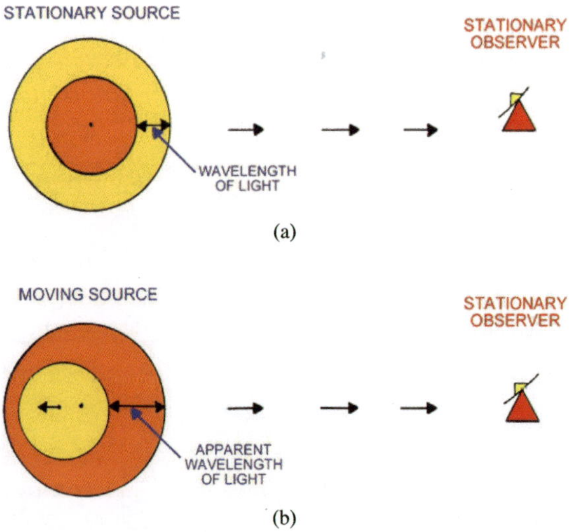

Figure 2.3. (a) Stationary light source and stationary observer. (b) Receding light source and stationary observer.

The change in wavelength of waves emitted by a moving source is considered a shift. The shift can be detected by examining the spectrum of light emitted by known elements. Many mechanisms at work in a star can radiate light. The mechanism of interest to us is the release of energy from an atom that has absorbed energy from the star. The energetically excited atom relaxes to a less energetic state by releasing energy in the form of an electromagnetic wave (light). Each kind of atom generates a spectrum, or a distribution of wavelengths, that is an identifying characteristic of that atom. For example, if we measure the spectra of hydrogen and helium in the laboratory, the spectrum of one hydrogen atom is the same as the spectrum of another hydrogen atom, but differs from the helium spectrum. By determining the spectra of atoms on Earth, we can identify the types of atoms present in stars by comparing their spectra.

The spectrum of electromagnetic radiation from a star can show dark lines or bands because some of the wavelengths of radiated light are absorbed as the light passes through a substance. The spectrum is known as an absorption line spectrum. The set of absorption lines is a signature for the substance.

We can compare the absorption line spectrum obtained in a laboratory with the spectrum of light emitted by the source. Figure 2.4 compares the laboratory absorption line spectrum, which we can consider a stationary observer, with the spectra associated with approaching and receding galaxies. The three dark absorption lines are shifted to the red (redshift) when the source is receding from the laboratory, while they are shifted to the blue (blueshift) when the source is approaching. Scientists measured absorption

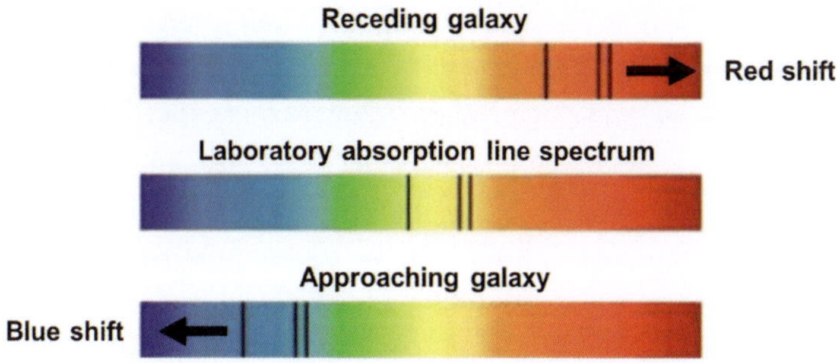

Figure 2.4. Doppler effect and absorption line spectrum.

line spectra of elements like hydrogen and helium and noticed that the absorption lines were redshifted, which implied the source was receding, or moving away, from Earth.

Cosmic Distance Ladder

Edwin Hubble depended on the measurement of astronomical distances to derive the velocity-distance relation for cosmological expansion. Several methods are used to measure astronomical distances because more than one method is needed to measure distances at all of the ranges encountered in astronomy. Some methods can measure distances to nearby astronomical objects, while other methods are needed to measure distances at intermediate to extreme distances. The methods build on each other and form a cosmic distance ladder where each rung of the ladder can measure greater distances than lower rungs. Two different methods can be calibrated when the range of applicability of the methods overlap. The methods are discussed here.

Distances to stars near the Earth are determined by triangulation: The distance to a point on a triangle is determined by measuring angles to the point from two other points at opposite ends of a baseline with a known length. In astronomy, the triangulation method known as the stellar parallax method uses the position of a nearby object such as a star viewed along two different lines of sight (Figure 2.5). The position of the nearby object appears to shift relative to a field of distant stars. A triangle is formed with the object at one apex and observations from two points of the Earth's orbit at the remaining two apexes. The greatest accuracy is achieved by maximizing the length of the baseline. The longest baseline is formed when observations are made on opposite sides of the Sun. The distance of the object from the Earth is calculated using the baseline length and the angles of measurements to the object.

The parallax method can estimate distances up to approximately 10,000 light-years. Measurements of larger distances are needed because we have instruments that observe light emanating from astronomical objects at distances stretching over billions of light-years. Light from these objects began its journey to the Earth billions of years ago. By observing distant astronomical objects, we are getting a glimpse of the universe as it appeared

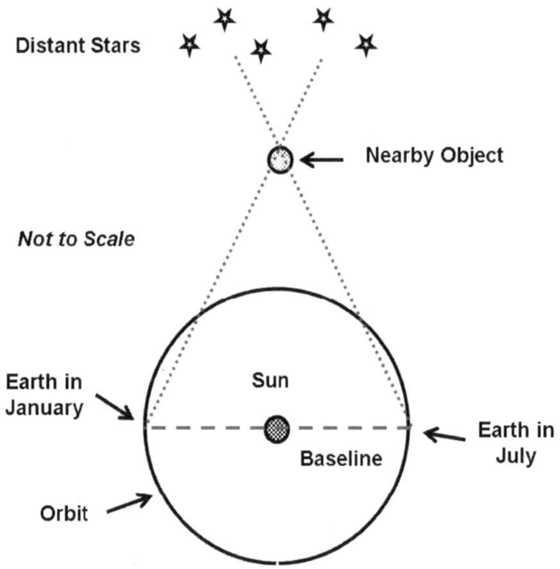

Figure 2.5. Stellar parallax.

billions of years ago, near the time of the Big Bang. How do we measure such vast distances?

One method, good to distances up to 100 million light-years, is based on a discovery by Henrietta Leavitt (1868–1921) at Harvard Observatory [Leavitt and Pickering, 1912; Leavitt, 1908]. She observed a pulsation in the visible light originating from some stars: the light dims and brightens periodically. These pulsating stars are known as Cepheid variables. The period of the pulsation is related to the luminosity, or total energy radiated each second, by a Cepheid variable. Thus, the period of pulsation is a "fingerprint" or "signature" of a Cepheid variable. If we determine the period of pulsation of a distant Cepheid variable, we can compare its observed luminosity to the luminosity we would expect based on studies of nearby Cepheid variables. The difference between the observed and expected luminosities of distant Cepheid variables is a measure of their distance from the Earth. Once we know the distance to a Cepheid variable, we have an estimate of the distance to its neighbors. Edwin Hubble used Cepheid variables to measure distances in his study of redshifted stellar spectra.

Cepheid variables are not well suited for measuring the distances of the most remote objects we can detect. Another astronomical discovery was needed, and it came in the form of the supernova. A supernova is a star that ends its life in a powerful and luminous explosion that can be seen over billions of light-years.

Supernovae may be classified as one of two types. A supernova that displays absorption lines associated with hydrogen is a Type II supernova, otherwise it is a Type I supernova. A Type II supernova occurs at the end of a massive star's lifetime. As massive stars age, the fusion process consumes available fuel and eventually reaches a stage when the repulsive pressure associated with nuclear fusion is overcome by gravitational forces associated with the mass of the star. The collapse of the core results in a Type II supernova.

For comparison, a Type I supernova can occur with less massive stars. An important example for the cosmic distance ladder is the Type Ia supernova. Suppose a carbon-oxygen white dwarf star is orbiting a companion star in a binary star system. If the gravitational attraction of the white dwarf star is enough to attract matter from its companion star, it can accumulate enough matter to increase its core temperature to the point where it initiates carbon fusion and results in an explosive runaway nuclear reaction. The resulting Type Ia supernova has a characteristic light curve that displays luminosity as a function of time.

The Type Ia supernova is an example of a standard candle. A standard candle in the astronomical sense is a light source with a known luminosity. By comparing a nearby standard candle such as a Cepheid variable star with the light curves of distant standard candles such as a Type Ia supernova, we can calibrate the Type 1a supernova as a standard candle for measuring great distances.

The methods described here illustrate methods used in a cosmic distance ladder.[1] The cosmic distance ladder is also known as the cosmological distance scale or the cosmic distance chain. The Stellar Parallax rung in Figure 2.6 uses methods that can measure astronomical distances up to 10,000 light-years, while the Cepheids rung measures astronomical distances between 1,000 light-years and 100 million light-years. The rung on the ladder named Distant Standards includes the use of Type Ia supernovae calibrated

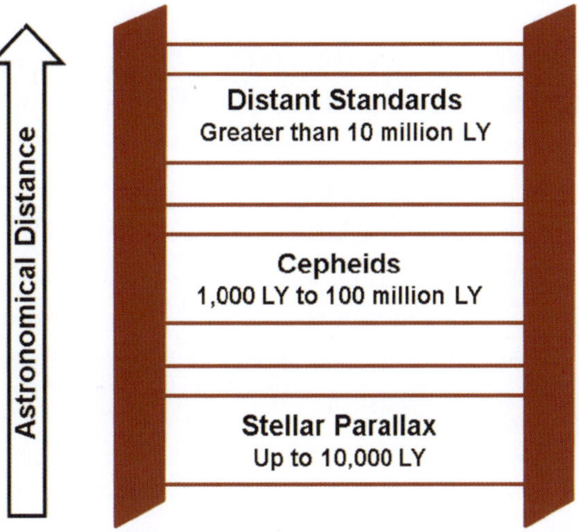

Figure 2.6. Illustration of a cosmic distance ladder (LY = light-years).

with Cepheid variables. It can measure astronomical distances ranging from 10 million light-years to billions of light-years.

Expansion of the Universe

In 1929, American astronomer Edwin Hubble and his colleagues were studying the properties of several galaxies. His colleague Vesto Slipher (1875–1969) observed a redshift in atomic spectra when compared with light from nearer, similar galaxies. Interpreting the redshifts as Doppler shifts, Hubble compared the galactic velocities relative to the Sun with their distances from the Sun and concluded that the galaxies were traveling away from the Earth.

Astronomers applied Hubble's methods to more distant galaxies and found that light from each of the galaxies was redshifted. They concluded that galaxies throughout the universe are receding from the Earth. Furthermore, they observed that recessional velocity is proportional to distance. The linear relationship between recessional velocity and distance is called Hubble's law and the slope of the straight line H_0 is known as Hubble's constant or the expansion rate of the universe.

Hubble's law says that more distant objects recede faster than less distant objects. This implies that the separation between galaxies is increasing. How can the apparent motion of galaxies be explained as a cosmological phenomenon?

If the universe is expanding, galaxies in the universe behave much like objects drawn on the surface of a balloon (Figure 2.7). The objects recede from each other as the balloon is inflated. The three objects in the figure move away from one another and the separation between the objects increases as the surface expands. Light emanating from galaxies in an expanding universe is redshifted because the galaxies are moving away from each other.

Hubble postulated that the universe should look the same from any point in the universe if it is expanding from a uniform point. Figure 2.8 shows how two observers, Observer A and Observer B, would see an expanding universe. The set of circles in the top row on the left side of the figure shows what Observer A would see at an early time t_1 and a later time t_2. The right side of the figure shows what Observer B would see. The locations of the observers are indicated by the yellow (light colored) circles. The bottom row of circles shows an overlay of the circles in the top row. The overlay is centered on the observer. Both observers see circles moving away from their location.

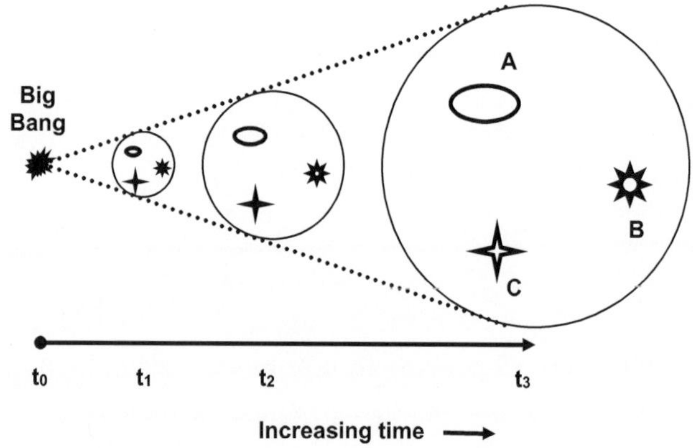

Figure 2.7. Illustration of the expanding universe.

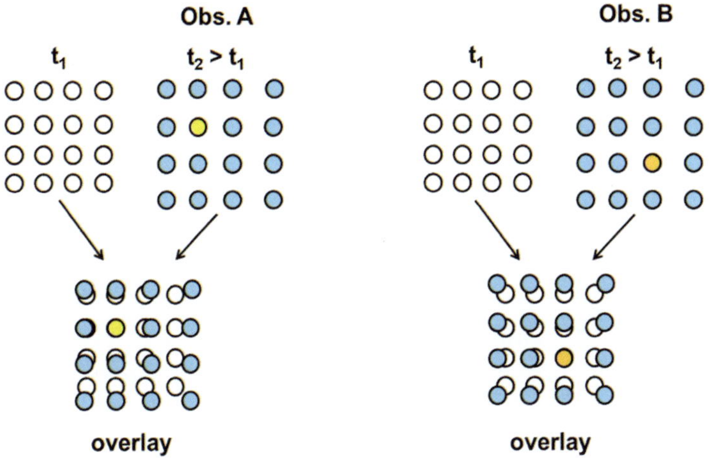

Figure 2.8. Every observer sees expansion [Krauss, 2012, pp. 11–14].

Point to Ponder: What Is Expanding?

Everywhere scientists looked in the sky, they saw receding galaxies. It appeared the universe was expanding, but what does that mean? The modern notion of an expanding universe was developed in the early 1900s. In his special theory of relativity, Albert Einstein [1905] replaced Isaac Newton's view [Newton, 1687] that time and space are absolute and independent with the idea that the three dimensions of space and one dimension of time were part of a four-dimensional spacetime continuum. The special theory of relativity applies only to nonaccelerating bodies. Einstein published a general theory of relativity a decade later [Einstein, 1915] that applied to accelerating bodies, such as a massive object moving in a gravitational field. According to the general theory of relativity, the universe expands as the spacetime continuum expands.

Einstein solved the equations of the general theory of relativity for a static, unchanging universe.[2] In 1922, Russian mathematician Alexander Friedmann (1888–1925) solved Einstein's equations for a model of a dynamic universe, that is, a universe that could change with time. Belgian cosmologist and priest Georges Lemaître (1894–1966) independently developed a dynamic universe model from Einstein's equations in 1927. Lemaître's model universe could be static, expand or contract. By 1931, Hubble's velocity-distance relation had been published and suggested the idea of an expanding universe. Lemaître proposed the hypothesis of the primeval atom in 1931 and used the primeval atom to initiate an explosive expansion of the universe. Lemaître was the first to propose the idea that the universe began in a moment of creation. By the 1940s, Russian-American astrophysicist George Gamow (1904–1968) applied Friedmann's and Lemaître's dynamic solutions of Einstein's equations to study the behavior of matter in the early universe. He predicted the existence of a cosmic background radiation that was not observed until the 1960s.

The term Big Bang was first used by astronomer Fred Hoyle (1915–2001) on a 1950 BBC radio broadcast to describe the apparently explosive expansion of the universe. Hoyle appeared to use the term derisively as he argued for a steady-state universe. [Singh, 2004, pp. 352–353] The steady-state theory was based on the hypothesis that matter was being continuously created in the expanding universe to preserve the homogeneity and isotropy of matter distributed throughout the universe. In this theory, the universe had no beginning and no end.

Point to Ponder: Speed of Recession

Hubble's law says that two objects separate from each other at a speed that exceeds the speed of light when the objects are far enough apart. It would appear that this violates one of the postulates made by Einstein [1905, p. 38] in his special theory of relativity. Einstein postulated that "light is always propagated in empty space with a definite velocity c which is independent of the state of motion of the emitting body." He argued that the speed of light is constant in empty space, and that matter cannot travel faster than the speed of light. Hubble's law and Einstein's postulate of the constancy of the speed of light are consistent with each other if we recognize that Hubble's law refers to the expansion of spacetime, while Einstein's postulate refers to the motion of an object within spacetime.

The modern steady-state theory was proposed by Hoyle [1948], Hermann Bondi, and Thomas Gold [Bondi and Gold, 1948]. Measurements of the distribution of matter have shown that the density of matter decreases as the universe expands. This observation and the observation of the Cosmic Microwave Background (CMB) Radiation discussed below do not support the steady-state universe hypothesis.

Cosmic Microwave Background Radiation

Hubble's law and an explosive origin of the universe suggest that we are moving away from the origin of an expanding universe (Figure 2.7). Suppose we imagine the expansion as a movie and roll it backward. If we trace the position of galaxies back in time, is it possible to find a time when all of the galaxies converge in both space and time? Would this be the origin of the universe?

Proponents of an expanding universe suggested that the expansion was a result of a primordial explosion. If such an explosion did take place, it is possible that debris from the explosion, some type of remnant, can still exist as evidence of the cataclysm?

CMB radiation was discovered in 1964 by Americans Arno A. Penzias (1933–) and Robert W. Wilson (1936–2013).[2] Penzias and Wilson were developing a sensitive instrument for measuring radiation having wavelengths on the order of seven to twenty centimeters (about three to eight inches). They opted to use this instrument — known as a horn reflector or microwave antenna — to observe any microwave radiation not originating on the Earth. Such cosmic radiation could interfere with satellite communications. Penzias and Wilson's employer, Bell Laboratories, wanted to determine if such interference existed.

Prior to 1964, no one had successfully managed to isolate the cosmic microwave frequencies in a study of radiation streaming to the Earth from outer space. Radiation from space covers a relatively broad range of spectral frequencies and can be loosely characterized as noise. Some of this noise could be inherent to the electrical device, it could arise as interference from Earth-based radiating systems such as television or radio transmitters, or it could originate in the cosmos. The challenge for Penzias and Wilson was to remove all sources of microwave radiation other than the cosmos.

When Penzias and Wilson performed their observations after a painstaking effort to eliminate all unwanted sources of radiation, they did not expect to find any signal. Instead, they detected a signal — a hiss — coming from every direction in which they aligned their antenna. The hiss was present everywhere in the sky and in all of the seasons. The energy of the radiation they observed was equivalent to an absolute temperature of about 3.5°K, or about –453°F. What is the meaning of this ubiquitous radiation?

Scientists[3] believe the radiation observed by Penzias and Wilson is the CMB radiation. It is a relic of the primordial explosion that initiated the cosmological redshifts that we attribute to an expanding universe. The ability to measure the cosmic microwave background has helped make cosmology an experimental science. Penzias and Wilson used a microwave receiver to discover CMB in 1965. The Cosmic Background Explorer (COBE) spacecraft was launched in November 1989 and was designed to improve our view of the early universe [NASA-COBE, 2009].

The COBE spacecraft operated for the four-year period from 1989 to 1993. Figure 2.9 shows the map of CMB temperature observations based

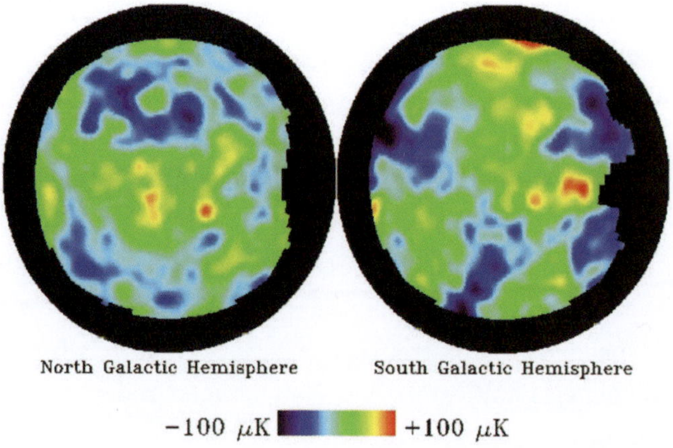

Figure 2.9. COBE-DMR map of CMB anisotropy [NASA-COBE, 2009].

on four years of data. The CMB displayed anisotropy for a temperature range of $\pm 100 \times 10^{-6}$°K. The CMB temperature fluctuations are presented in the figure as the difference between the 2.73°K average temperature of the radiation field and the observed value. Temperature fluctuations range from approximately -100×10^{-6}°K in blue to $+100 \times 10^{-6}$°K in red. The data were measured by a differential microwave radiometer. Observed CMB variations show the presence of anisotropy in the distribution of matter and energy when the Universe was approximately 380,000 years old.

The Wilkinson Microwave Anisotropy Probe (WMAP)[4] provided more accurate measurements of the anisotropy observed by COBE. The WMAP spacecraft was launched in 2001 [NASA-WMAP, 2010]. It was originally named MAP and was later renamed WMAP to honor David T. Wilkinson, a Princeton University cosmologist and a founding team member who died in September 2002. WMAP observed five different microwave frequencies in the full sky. Figure 2.10 shows CMB temperature anisotropy when the universe was approximately 375,000 years old. The anisotropy map is based on 9 years of data and shows a temperature range of $\pm 200 \times 10^{-6}$°K. relative to an average temperature of 2.73°K. The WMAP mission determined that the universe was 13.77 billion years old with an error of less than 1%.

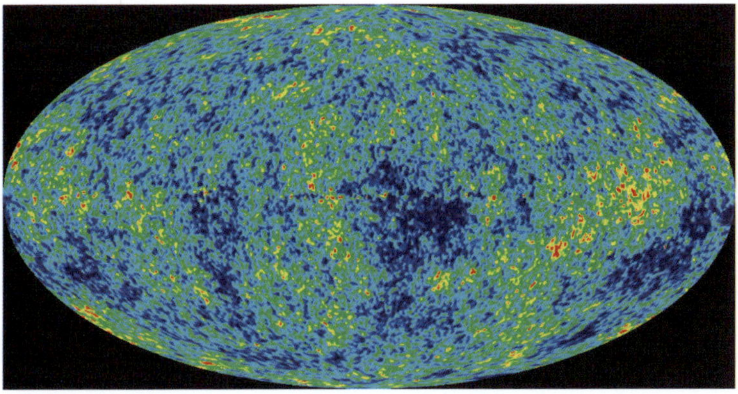

Figure 2.10. WMAP of CMB temperature anisotropy [NASA-WMAP, 2010].

Point to Ponder: Hubble Constant Discrepancy

Hubble [1929] first estimated the expansion rate of the universe H_0 to be 500 km/s/Mpc Hubble and Humason [1931] included more distant galaxies and estimated the expansion rate H_0 to be 560 km/s/Mpc. The inclusion of more distant objects led to a significant decrease in H_0 since the middle of the twentieth century.

Two modern estimates of H_0 have been provided by Riess *et al.* [2016] and the Planck collaboration [2020]. Riess *et al.* [2016] used the Hubble Space Telescope to observe Cepheid variables and estimated H_0 to be 73.4 km/s/Mpc with uncertainty ±1.74 km/s/Mpc. The Planck collaboration [2020] used cosmic microwave background (CMB) data from the European Space Agency's Planck satellite to estimate H_0 to be 67.4 km/s/Mpc with uncertainty ±0.5 km/s/Mpc. These two estimates are based on independent methods and are relatively close, but their values with associated uncertainties do not overlap.

The discrepancy in H_0 values is an example of an anomaly in normal science.

Observational evidence supports the idea that the universe appears to be expanding as astronomical objects move away from one another. The universe is approximately 13.8 billion years old, and we have observed a CMB temperature of approximately 2.73°K that appears to be a relic of the Big Bang.

Chapter 3

Evidence for Inflationary Cosmology

The previous chapter provided evidence for an expanding universe. Scientists believe that the rate of expansion immediately after the Big Bang was very large for a short period of time before the expansion rate slowed to the rate of expansion discussed in Chapter 2. The very large rate of expansion is known as inflation and is a distinguishing feature of inflationary cosmology. To better appreciate the meaning of inflationary cosmology, we review the development of cosmological models in the early 1900s. We then present evidence for the existence of dark matter and dark energy and begin to explore their cosmological implications. We begin with an introduction to classical physics.

Classical Mechanics and Electrodynamics

Classical physics applies to many of the natural systems we encounter in everyday life.[1] These systems are often referred to as macroscopic systems because we can observe them with our unaided senses. Smaller-scale systems, like the behavior of electrons in atoms, are called microscopic systems and do not always obey the "laws" of classical physics. They obey the "laws" of quantum physics and are discussed later.

Isaac Newton's *Principia*, or *Mathematical Principles of Natural Philosophy*, was the first "*Bible*" of classical physics [Newton, 1687]. Newton introduced a theory capable of precisely describing the motion of macroscopic objects. He assumed that "Absolute, true, and mathematical time, of itself, and from its own nature, flows equably without relation to anything external… Absolute space, in its own nature, without relation to anything external, remains always similar and immovable" [*Ibid.*, 1687, p. 6]. According to Newton, time and space are independent of each other, and are permanent. A belief in the permanence of time and space implies the existence of points in space and instants of time independent of objects or observers confined within the space and time domains. From this perspective, an observer moving at a very great speed would measure the passage of time at the same rate as a motionless observer.

Building on his view of space and time, Newton formulated three "Laws" to describe the motion of objects [Newton, 1687, p. 13]:

- Law I: "Every body continues in its state of rest, or of uniform motion in a right line, unless it is compelled to change that state by forces impressed upon it." Law I contradicts Aristotle's view that the goal of every object is to attain rest and unchangeability.
- Law II: "The change of motion is proportional to the motive force impressed; and is made in the direction of the right line in which that force is impressed ..."
- Law III: "To every action there is always opposed an equal reaction ..."

Newton's notions of space and time and his three Laws are the essential elements of the theory known as nonrelativistic classical mechanics. This theory epitomizes a deterministic view of nature. If we can measure the initial motion (or momentum) of an object and its initial location simultaneously, as Newton's theory allows, then we can, in principle, calculate the historic and future behavior of the object provided we know all of the forces acting upon the object during the time of interest to us. What are the fundamental forces of nature?

Until the twentieth century, only two fundamental forces were known: the gravitational force and the electromagnetic force. These two forces were sufficient to describe the observed behavior of all objects. Every massive object experiences the pull of gravity. The motion of astronomical bodies is governed by the gravitational force or gravity. Electricity and light are examples of electromagnetism. Electromagnetism is associated with objects that have an electrical charge, such as electrons. It is the force that binds electrons to atomic nuclei. No other forces were needed for pre-twentieth century science.

As a prelude to understanding the modern extension of classical Newtonian mechanics to relativistic speeds, we need to learn more about the electromagnetic interaction. Electric and magnetic phenomena are special cases of a more general class of phenomena known as electromagnetic phenomena. One of the first and most thorough experimental studies of electromagnetic phenomena was performed by Englishman Michael Faraday (1791–1867) in the nineteenth century. Faraday's work provided much of the experimental data used by his contemporaries to construct a mathematical theory of electromagnetism.

The most successful mathematical formulation was presented by Scottish mathematician and physicist James Clerk Maxwell (1831–1879). Maxwell, building on the work of other researchers, including Faraday, discovered a set of equations which could be used to describe all of the then-known electromagnetic observations. Maxwell's equations synthesized experimental observations in electricity and magnetism, and showed that electrical and magnetic forces are different manifestations of a single electromagnetic force. In addition, Maxwell used his equations to predict a large, but finite, speed for light. His prediction was later confirmed by experiment. Maxwell's electromagnetic equations are comparable in importance to Newton's Laws of mechanics.

The combination of Newton's mechanics and Maxwell's equations form the theoretical basis of the subject known as classical electrodynamics: the study of the motion of electrically charged objects in gravitational and electromagnetic fields. Classical electrodynamics, and especially the speed of light, played a major role in Einstein's development of relativity.

Einstein's Relativity

Experimentalists had discovered that the speed of light was the same whether you were measuring it from a reference frame at rest or while moving at a constant speed not exceeding the speed of light. A reference frame consists of a coordinate system that can be used to mathematically describe observations of physical phenomena. Indeed, it appears to be physically impossible for any object to move faster than the speed of light. Albert Einstein (1879–1955) presented his explanation of this phenomenon[2] in his 1905 paper entitled "On the Electrodynamics of Moving Bodies" [Einstein, 1905a].

Before Einstein's paper, physicists had devised a set of rules for transforming the description of the motion of an object from one uniformly moving (nonaccelerating) reference frame to another. The transformation rules treated the dimensions of space and time as mathematically independent and did not change the form of Newton's Laws. Einstein proposed a new set of transformation rules, which made the form of Maxwell's equations independent of the uniform motion of an observer. These new transformation rules, known collectively as the Lorentz transformation after their founder, Dutch physicist Henrik Anton Lorentz (1853–1928), calculated a speed of light that did not change from one uniformly moving reference frame to another. The Lorentz transformation introduced a relationship between space and time measurements that resulted in a modification of Newton's laws of mechanics. Einstein rejected Newton's concept of "absolute time" when he adopted the Lorentz transformation.

In addition to replacing Newton's "absolute time" with relative time, Einstein changed the physics community's understanding of simultaneity. He showed that we should regard the notion of simultaneity as a concept that depends on the motion of the observer.

Point to Ponder: Do We Need a Reference Frame to Understand Duration?

Newton's concepts of space and time, though still useful at speeds much less than the speed of light, are no longer considered fundamental. Furthermore, it became meaningless to specify the duration of an event without specifying the reference frame in which the duration was measured. Thus, to claim an event lasted one day is no better than saying the same event lasted millions of years if the reference frame of the observer is not known. This point provides one means of reconciling the meaning of time in the Genesis version of creation with the use of time in the secular view, as we discuss later.

The theory presented by Einstein [1905a] is known today as the "special theory of relativity." It is concerned with reference frames moving uniformly with respect to other reference frames. These uniformly moving reference frames are known as nonaccelerating, or inertial, reference frames. As a part of the theory of relativity, Einstein showed that mass and energy are but different manifestations of the same physical quantity. He said that mass can be viewed as a measure of the energy content of an object [Einstein, 1905b]. The equivalence of mass and energy is the basis of nuclear power and nuclear weapons.

Einstein was unable to present a theory of relativity in 1905 for reference frames which were accelerating with respect to one another. After a decade of research, Einstein [1916] published a relativity theory for accelerating reference frames. The 1916 theory was called the "general theory of relativity." A year later Einstein [1917] published an article that explored cosmological models within the context of the general theory of relativity.

The general theory provides a geometric interpretation of the gravitational interaction. Simply stated, the presence of matter curves spacetime, and the curvature of spacetime affects the trajectory of matter.[3] The general theory predicts, among other things, the existence of gravity waves and black holes. Especially relevant to us in our search for a contemporary worldview are the cosmological models arising from the general theory.

Point to Ponder: Was Einstein Original?

Bjerknes [2003] presented a controversial argument that Einstein took more credit than he deserved for developing relativity. What makes Bjerknes' argument notable is his explicit use of the original scientific literature to support his thesis that Einstein plagiarized some of his most important ideas. An alternative view is that Einstein built on the work of others, which is typical of the scientific process, but did not properly reference their work. Stott [2012] described how Charles Darwin (1809–1882) did not properly reference the work of those who came before him. A third view is that all of Einstein's ideas were original. Historians of science will have to decide if Einstein was ethical and if he deserves the credit he has received.

Open and Closed Universes

Many published cosmological models are based on Einstein's general theory of relativity.[4] The models can be separated into essentially two distinct

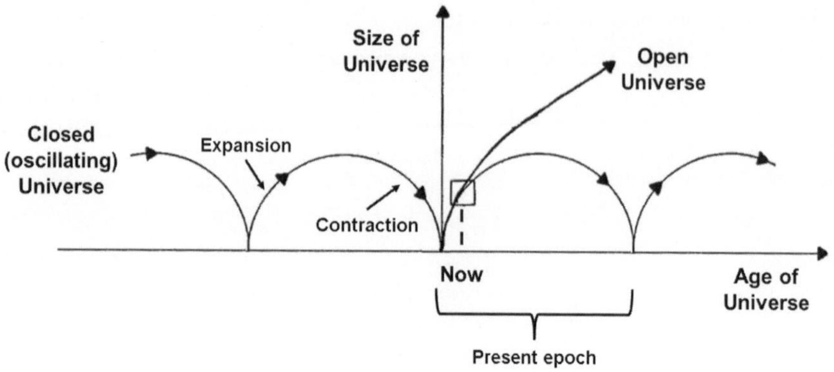

Figure 3.1. Open universe and closed universe models.

classes: open universe models, and closed universe models. Illustrations of the open universe and closed universe models are presented in Figure 3.1.

Open universe models assume the universe began as an event in spacetime. The initial event was the release of all the mass-energy of the universe. Models of this type have been dubbed "Big Bang" models. Following the initial event, the spacetime domain containing the mass-energy of the universe begins to expand. Open universe models assume the expansion of spacetime will continue indefinitely following the initial event.

By contrast, closed universe models assume the expansion of the spacetime domain will eventually stop. After the expansion stops, spacetime begins to contract until all of the mass-energy of the universe is once again compressed to a single spacetime point. Having returned to a single point, another release of mass-energy may occur, though how is an unanswered question. If a release occurs, spacetime once again starts to expand. If we assume the expansion and contraction phases are repeated periodically, we obtain a closed universe model called the oscillating universe. Oscillating universe models are cyclic in character, where the cycle is one of expansion and then contraction of spacetime.

Dark Matter

Matter has historically been viewed as a physical object with mass. The concept of matter is an ambiguous concept because of the relativistic equivalence between mass and energy discovered by Einstein in 1905.

Mass can be transformed into energy, and energy can be transformed into mass.

The luminous matter we see in the universe represents a class of matter called ordinary matter. Observations of the distribution and motion of luminous mass in the universe implied the need for a new form of matter.[5] California Institute of Technology astrophysicist Fritz Zwicky (1898–1974) found that there was not enough observable matter in the Coma cluster of galaxies to account for the motion of galaxies in the cluster [Zwicky, 1933, 1937]. Rather than reject accepted theories of gravitation, scientists concluded that there must be matter that had not yet been observed. Zwicky called the missing matter "dark matter."

American astrophysicist Vera Rubin (1928–2016) and colleagues confirmed Zwicky's observations of dark matter in the 1960s and1970s by studying the motion of spiral galaxies [Bennet *et al.*, 2018]. Rubin noticed that stars at the fringes of the Andromeda Galaxy moved faster than expected based on the gravitational attraction associated with the mass of luminous stars in the galaxy.

Working with colleague Kent Ford, Rubin [Rubin *et al.*, 1976a, 1976b] measured "orbital speeds of hydrogen gas clouds in many other spiral galaxies (by studying Doppler shifts in the spectra of hydrogen gas) and discovered that the behavior seen in Andromeda is common. Although Rubin and Ford did not immediately recognize the significance of the results, they were soon arguing that the universe must contain substantial quantities of dark matter" [Bennet *et al.*, 2018, p. 472]. By the 1980s, the evidence for dark matter had convinced the scientific community that our best theory of gravity was wrong, or large amounts of dark matter must exist.

Dark matter and ordinary matter are present in approximately the same regions of space. Conversely, dark matter does not seem to be present in regions of space without visible galaxies. Both ordinary matter and dark matter interact gravitationally. We can detect ordinary matter when it interacts with electromagnetic radiation (light). Dark matter seems to interact only weakly or not at all with electromagnetic radiation. Some physicists have postulated that dark matter is a new type of particle called a WIMP (weakly interacting massive particle) or astronomical object called a MACHO (massive compact halo object).

Dark matter is matter we cannot see. The existence of dark matter is implied by observing its apparent gravitational influence on the distribution of ordinary matter. These observations can be explained if matter consists of approximately 15–20% ordinary matter and 80–85% dark matter. Dark matter remains a mystery to science and its existence implies that most of the matter in the universe is a mystery. Even our understanding of ordinary matter is continuing to evolve, as we discuss in Chapter 4.

Dark Energy and Inflation

In 1998, two independent teams of astronomers were trying to extend Edwin Hubble's law of universal expansion by measuring the distance to supernovae in galaxies with larger redshifts than observed previously. Figure 3.2 is a sketch of Hubble's law. The solid line illustrates the linear relationship associated with relatively small redshift galaxies. The three broken lines (dots and dashes) at higher redshift values represent the part of the relationship that needed to be measured. The two teams thought they would observe a decelerating expansion. Instead, they found that the expansion of the universe followed the accelerating expansion curve represented by a dashed line. According to the observations, the universe is expanding at an increasing rate. What is causing this accelerating expansion?

The accelerating expansion is attributed to a form of energy called dark energy. When viewed within the context of general relativity, dark energy appears to behave as a gravitationally repulsive form of energy that exerts a negative pressure.

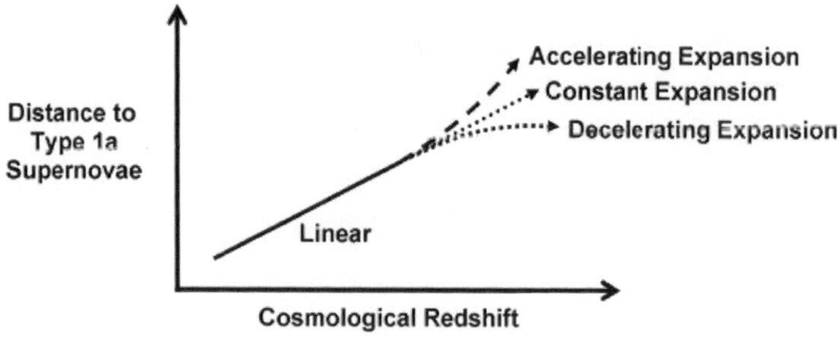

Figure 3.2. Extension of Hubble's law.

Dark energy is included in cosmological models by adding a term called the cosmological constant to a general relativistic equation that relates matter and the geometry of space-time. Einstein first introduced the cosmological constant in 1917 in an effort to explain a universe which was then considered unchanging, only to learn later from Hubble's measurements that the universe appeared to be expanding. Einstein viewed the cosmological constant as a major blunder,[7] but the cosmological constant has been resurrected to help explain observations that the expansion rate of the universe is not constant. The cosmological constant is often represented by the Greek letter lambda Λ. A cosmology that includes dark energy is discussed in Chapter 5.

Dark energy is represented by the cosmological constant if dark energy is constant in time and space, while dark energy is represented by a variable called quintessence if dark energy is a function of time and space. The term quintessence means fifth essence and refers to the fifth element of the ancient Greeks, who thought that the elements were earth, air, fire, water, and aether. Aether was the fifth element that filled the universe beyond Earth. The spacetime dependence appears in an equation of state that relates pressure and density.

Dark energy represents approximately 70% of the cosmic energy density. Ordinary matter (~5%) and dark matter (~25%) contribute the rest. The percentages are subject to revision as research progresses. Ordinary matter is the building material for all of the astronomical objects we can observe, such as gas clouds, stars, and planets. Very little is known about dark energy. If we combine dark energy and dark matter, we find that approximately 95% of the mass-energy in the universe is poorly understood. Research to understand dark matter and dark energy is continuing.[5]

Cosmological Model Parameters

Astronomical observations indicate that the universe is expanding as a result of a "Big Bang," or primordial explosion. Big Bang models are based on measurements of astronomical distances, observations of the cosmological redshift, the cosmic background radiation, and the distribution of matter in the universe.[8]

Modern cosmological models are based on the general theory of relativity. Two important assumptions are made to simplify the mathematics:

1. Homogeneity Assumption: Matter in the universe is uniformly distributed.
2. Isotropy Assumption: The distribution of matter in the universe does not depend on direction.

These two assumptions are subject to experimental testing and are combined in the principle known as the Cosmological Principle: the distribution of matter in the universe is homogeneous and isotropic.

Measurements of the cosmic background radiation are used to determine the isotropy of matter in the universe. The distribution of matter in the universe is isotropic if the distribution does not depend on which direction we look. If there is more matter in one direction than another, the universe is anisotropic. The isotropy of the universe is one of the fundamental parameters of cosmological models that can be measured. The standard cosmological model depends on three observable parameters: the Hubble expansion rate; a dimensionless deceleration parameter; and the critical density of the universe. We discussed the Hubble expansion rate previously. The deceleration parameter is estimated from the relationship between the magnitude of the redshifts of E-type galaxies and their distances from the Earth. E-type galaxies are standard candles.

The third cosmological parameter is the critical density of the universe. Density in a cosmological model is a measure of the matter contained in the universe. If we can estimate the density of the universe, we can decide whether the density measurement supports an open, closed, or flat universe model. The density of a flat universe is the critical density, or third cosmological parameter.

We must estimate the density of the universe to determine the type of universe we live in. The observed density is the observable matter of the universe divided by the estimated volume of the universe. The critical density is the density of the flat universe shown in Figure 3.3. The ratio of observed density to critical density is often represented by the symbol Ω_0 (called omega zero). If Ω_0 is greater than 1, $\Omega_0 > 1$, the universe is closed. On the other hand, if Ω_0 is less than 1, $\Omega_0 < 1$, the universe is open and should

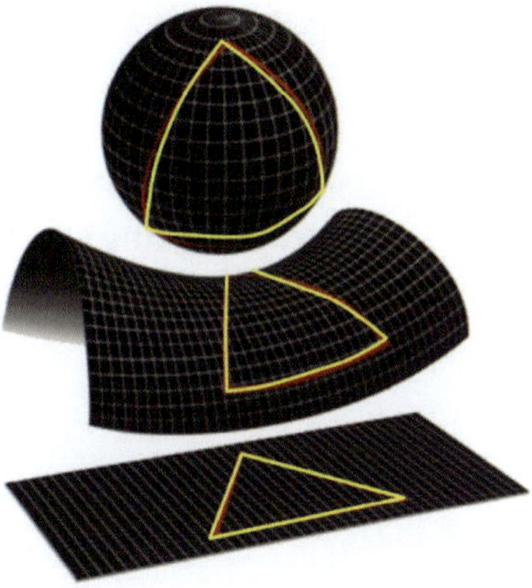

Figure 3.3. Geometry of possible universes [NASA Geometry, 2020].

continue to expand indefinitely. If the observed density equals the critical density, then $\Omega_0 = 1$ and the universe is flat. To decide, we must estimate Ω_0. The different geometries are illustrated in Figure 3.3.

Answers to two problems must be found: how much matter is in the universe, and how big is the universe? A somewhat reliable estimate of the size of the universe is obtained by using the cosmological redshift to determine the rate of expansion of the universe. With this information, we use the speed of light — the greatest known speed of any physical object — to estimate the radius of the universe. The size of the universe can then be estimated for a given geometry.

The amount of matter in the universe is estimated by simply adding up all of the matter we observe. The amount of matter in astronomical objects such as galaxies and nebulae is determined by studying the motion of astronomical candles (such as quasars or star clusters) within the object. Evaluating the motion of an astronomical candle in terms of gravitational laws leads to an estimate of the matter of the object. We must correct for the amount of matter we cannot observe due to the limitations of our instruments. Dividing the corrected amount of matter in the universe by the size of the universe gives us an observed density to compare with the critical density of our cosmological models. There are a number of uncertainties and assumptions built into the procedure outlined above. For example, what if we do not know the correct rate of expansion of the universe? Has the rate of expansion changed?

Inflation and Expansion Rate

Inflation is an example that shows the expansion rate has changed with time.[9] The cosmological constant can be used to represent inflation in cosmological models. It is added to the general relativistic equation that relates matter and the geometry of space-time. A cosmology that includes inflation is discussed in Chapter 5.

Figure 3.4 illustrates an inflationary period during the expansion of the universe when the expansion rate was greater than the rate measured by Hubble. There is a period of rapid expansion immediately after the Big Bang. The "Old Size" refers to the size of the universe without inflation. The inflationary expansion yields a bigger universe than the expansion without inflation.

Another source of error in our calculation of the universal density is the estimate of the amount of matter in the universe. Estimates of universal matter ordinarily invoke the Cosmological Principle: the distribution of matter in the universe is uniform (homogenous) and independent of the direction in which we look (isotropic). If we look at our solar system, at our Milky Way galaxy, or even at the Local Group of galaxies, we do not see a uniform distribution of matter. The Cosmological Principle does not apply on this scale because the gravitational influence of astronomical objects on one another in the Local Group disrupts the expected uniformity. We must

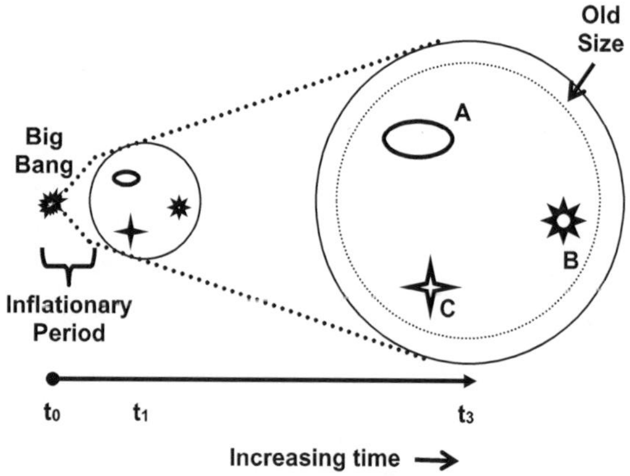

Figure 3.4. Inflation (not to scale).

look at an even larger region of space before we begin to observe a uniform and homogeneous distribution of matter. For the region of the universe we can observe, the Cosmological Principle is reasonable. What about elsewhere in the universe? Does the Cosmological Principle actually apply everywhere? The search for answers to questions like these has been the subject of research by NASA [NASA-WMAP Science Team, 2011].

One possible source of error in our estimate of the size of the universe is our assumption that we have accurately measured the masses of particles populating the universe. This assumption has been challenged by the observation that the neutrino has mass. The neutrino was originally introduced into physics to save the laws of conservation of energy and momentum. The neutrino in the standard model of particle physics discussed in Chapter 4 is supposed to be massless. A massless particle like the photon is essentially a quantum of energy. According to Einstein's special theory of relativity, massless particles can only travel at the speed of light. Particles with a rest mass — a nonzero value of mass as measured by an observer moving with the particle — move at speeds less than the speed of light, but never equal to or are greater than light speed. Some people have suggested massive particles can travel faster than the speed of light, but such particles, which are called tachyons, have never been observed. It now appears that the neutrino may actually have a small but nonzero rest mass. The abundance of massive neutrinos in the universe could contribute to the density of the universe, but research to date suggests that the mass of the neutrino is too small to account for all of the dark matter that has been inferred.

Black holes are another possible source of mass that may not be adequately accounted for in the calculation of the universal density. Black holes are very difficult to detect. According to theory, the gravitational field of a black hole is so strong that light cannot escape by classical methods from its gravitational pull. A quantum mechanical phenomenon known as tunneling is a means by which particles can escape black holes, and provides a mechanism for detecting them. Several astronomical objects have many of the properties people expect a black hole to have.[10] We do not know how many black holes there are. Consequently, their mass, which could be substantial, may be underestimated.

Age of the Universe

The determination of parameters used in cosmological models is not exact enough to definitely choose the model that applies to our universe. Parameter estimates give us evidence for judging the quality of competing models. For example, estimates of the age of the universe presently favor an age of approximately 13.8 billion years. This is very different from the estimate derived by Irish Archbishop James Ussher (1581–1656) in 1650 using Biblical chronology.[6] Ussher used the genealogy of Biblical characters to estimate that the earth was created around 4004 BCE. This would make the current age of the earth about 6,000 years old.

The age of the universe can be a point of contention between secularists and nonsecularists. Even within the scientific community, the age of the universe is a question worthy of debate. Scientists believe that the age of the universe is measured in billions of years, and the question that remains is how many billions of years. This age refers to the age of the universe measured from the Big Bang. An oscillating universe model would be ageless.

Personal Anecdote: Age of the Universe

I used a variation of the general theory of relativity to publish an estimate of the age of the universe in the mainstream journal *Physical Review A* in 1988 [Fanchi, 1988]. At the time, the Hubble expansion rate yielded an age of the universe between 10 and 20 billion years. Now scientists believe that inflation was an important event early in the life of the universe, even though questions remain about how inflation occurred, and that the Hubble expansion rate yields an age of 13.8 billion years subject to the Hubble constant discrepancy discussed in Chapter 2. Furthermore, data supporting inflation imply that the universal expansion rate may have varied over the life of the universe. This is an example of how scientific knowledge can be refined, change, and remain subject to further change because of factors that are not understood very well.

Scattering

Figure 3.4 illustrates the scientific view that the origin of the universe occurred prior to inflation. A scientific understanding of the Big Bang depends on our understanding of the subatomic world. One of the best methods for exploring the subatomic realm and gaining insight into the Big Bang is scattering.

Scattering is a simple concept with many everyday examples.[1] One familiar example is the case of a ball striking a bat. The collision between a ball and a bat causes a change in the initial direction of the ball. This change in direction of a moving object due to an interaction or collision with another object is an example of scattering.

Scattering experiments are used to study the behavior of subatomic systems. The machines needed to perform these experiments tend to be large and expensive. The chief function of scattering machines is to guide a collimated beam of particles into either a stationary material target or into another beam of particles. It is usually necessary to accelerate the incident beam of particles until the particles have attained a desired velocity. For this reason, machines built to perform particle scattering experiments are called particle accelerators. Particle accelerators are used to recreate on a small scale the conditions that may have existed when the universe was very small. If we want to understand the first few moments of the Big Bang, we need to understand high energy collisions between particles.

Some examples of processes and interactions which have been observed during scattering experiments are represented by Feynman diagrams, which were introduced in 1948 by American physicist Richard Feynman (1918–1988). Figure 3.5 shows a few examples of Feynman diagrams. The top diagram illustrates the notation. The horizontal axis represents one space dimension and the vertical axis represents the time dimension. A point on the diagram is given by a space coordinate and a time coordinate and is known as an event. An arrow represents a particle, and a dot, or vertex, denotes the spacetime event when an interaction occurs between two or more particles.

Pair production and annihilation are shown on the left-hand side of Figure 3.5. The wavy line denotes a photon, or particle of light. The symbols

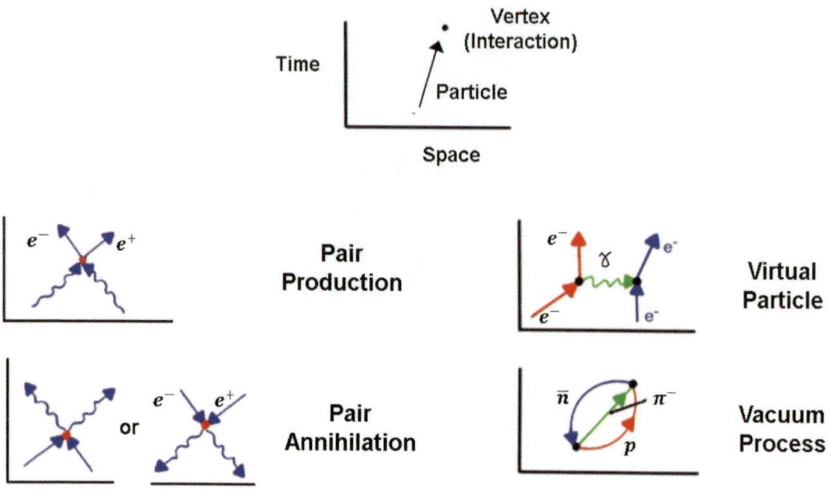

Figure 3.5. Examples of particle interactions.

e^- and e^+ denote an electron and its antiparticle, the positron, respectively. A particle and its antiparticle are examples of matter and antimatter.

Antimatter particles have the same mass as matter particles. They differ in properties such as electric charge: the charge of an antimatter particle is opposite that of a matter particle. An example of a matter–antimatter pair is the negatively charged electron e^- and positively charged positron e^+. Matter and antimatter are produced in the pair production process as a matter–antimatter pair when sufficient electromagnetic energy is available. When a matter–antimatter pair interact, they annihilate each other and transform into electromagnetic radiation.

The "virtual particle" diagram on the right-hand side of Figure 3.5 shows what may happen when two electrons pass near each other. Historically, the two electrons are said to repel each other because they have the same electrical charge. The repulsion was caused by one electron interacting with the electric field generated by the other electron. This point of view is based on the concept of action at a distance, namely, that two objects separated in space interact with one another when one object interacts with the field generated by the other object.

On a more fundamental level, we say the electrons exchange a photon γ, or quantum of light. A quantum is a packet of energy. The photon exists for a very short time and is consequently called a virtual particle. It is the interaction with the virtual photon which causes the change in trajectory, or scattering, of the two electrons. The latter view, namely the exchange of a particle, is a local interaction in which the exchanged particle cannot exceed the speed of light. The local concept of interactions and the concept of action at a distance are competing paradigms.

Interaction by action at a distance is an example of a nonlocal interaction. Locality and nonlocality refer to the correlation of events with spacelike separation. The spacetime diagram shown in Figure 3.6 illustrates regions of spacelike, timelike, and lightlike separation. The figure shows two space axes and a time axis. The axes define a reference frame, or a coordinate system, that can be used to represent the trajectory of an object in spacetime. An event is a point on the spacetime diagram. Time is measured along the time axis from past in the lower half of the diagram to future in the upper half of the diagram. The present moment is the event at the origin of the axes. The worldline traces the path of events.

Two events separating from each other at the speed of light are events on the lightlike boundaries of the two cones in the figure. The cones are called light cones, with the upper cone in the future and the lower cone in the past. The point where the tips of the cones meet is the present. If two events are separating at a speed less than the speed of light, they are in the timelike region of the light cones. If two events are separating at a speed greater than the speed of light, they are in the spacelike region outside of the

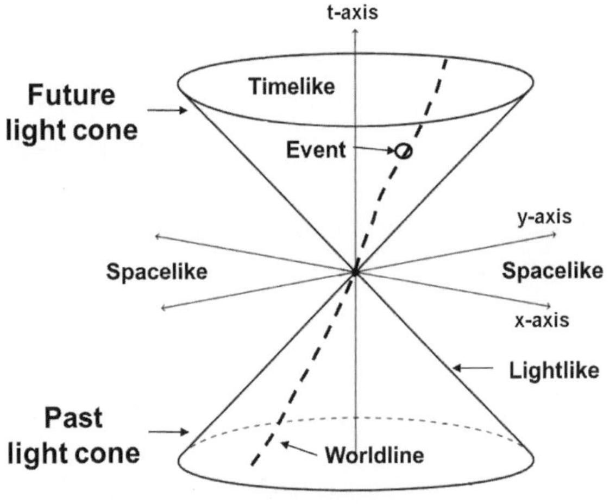

Figure 3.6. Spacetime diagram.

light cones. Given the notions of timelike, spacelike, and lightlike separation of events, a local theory does not allow the existence of greater than light speed correlations between events, while a nonlocal theory does.

Point to Ponder: Can We Observe Spacelike Separated Galaxies?

Astronomical observations and Hubble's law have shown that galaxies can be separated by spacelike distances. According to the special theory of relativity, information cannot be shared between spacelike separated objects because information cannot travel faster than the speed of light. This implies that galaxies beyond our own will eventually disappear if the universe continues to expand because they will be too far away for their light to reach us.

Regions that are spacelike separated can be considered isolated systems because they cannot share information. The speed of light establishes a horizon that restricts communication between spacelike separated regions of the universe. This observation is known as the Horizon Problem.

An especially relevant process for our purposes is the vacuum process shown in Figure 3.5. The example shows the production, interaction, and annihilation of a proton (p) with positive electric charge, an antineutron ($\overline{n}$), and a negatively charged pion (π^-). The vacuum process represents the physical situation in which virtual particles arise from the vacuum, interact, and then return to the vacuum all within a short time duration satisfying German physicist Werner Heisenberg's uncertainty principle. If we want to understand the vacuum process, which has been proposed as a natural mechanism for initiating the origin of the universe, we need to understand the uncertainty principle[2–3] and its role in the transition from classical to quantum physics. These topics are discussed in Chapter 4.

❁ ❁ ❁

We have introduced evidence for dark matter, dark energy, and inflation. Each of these topics can have a significant impact on the performance of cosmological models, including forecasts of the future of the universe. We have shown that dark matter and dark energy are much more abundant than ordinary matter. To better appreciate the scientific view of the origin and evolution of the universe, we discuss the leading description of ordinary matter in the next chapter.

Chapter 4

The Big Bang

The scientific worldview is a working theory of the origin and evolution of the physical universe, and the origin and evolution of life. It is natural to begin our description of the origin of the universe at the beginning. Yet this implies there was a beginning: an implication that has theological implications and is not required by the best observations available today. Was there a beginning? How can we decide if the universe had an origin? To understand the scientific explanation of the origin and evolution of the universe, we need to understand how matter behaves on the subatomic scale. The most widely accepted theory of the subatomic world is the standard model of particle physics. It is used to suggest a mechanism for initiating the Big Bang.

Transition from Classical Physics to Quantum Physics

A revolution in physics was born at the beginning of the twentieth century. Prior to 1900 scientists were able to use classical physics to explain virtually all known experimental measurements. Only the observations of a few experiments seemed to withstand the concerted attack of nineteenth-century theorists. Yet an eventual understanding of these experiments required the reevaluation and replacement of many previously unchallenged concepts. The constancy of the speed of light in a vacuum, for example, was understood only when Albert Einstein (1879–1955) rejected Isaac Newton's (1643–1727) view of absolute and independent space and time. Another experimental observation, the radiation emitted by heated black bodies, was explained by the rejection of the previously established concept that energy was continuous. The black body experiment, and the physical theory emerging from it, are now considered.

Black Body Radiation

Our understanding of black body radiation and its scientific significance can be enhanced by considering the wave description of light. Maxwell's equations in classical physics describe electromagnetic phenomena and can be used to show that light has many of the characteristics of water waves (Figure 4.1).

The length of a wave from one point on the wave to an equivalent point is called the wavelength. The number of waves passing a particular point in a specified time interval is the frequency of the wave. A succession of similar waves is a wave train, and a localized wave that travels by itself is a wave packet. The water wave on the left-hand side of Figure 4.1 is a wave packet in the figure and is one wave in a wave train. The mathematical equation for describing the motion of a wave has the same form as the equation for describing the motion of light.

The wavelengths of light we see, visible light, represent only a very narrow band of wavelengths of the more general phenomenon known as electromagnetic radiation. Radio waves are an example of electromagnetic radiation with relatively long wavelengths. Visible light wavelengths are thousands of times shorter than the wavelengths of radio waves. Yet electromagnetic radiation with wavelengths shorter than the wavelength of visible light exists. The wavelength of X-rays, for instance, is one hundred to one thousand times smaller than the wavelength of visible light. Gamma rays, products of nuclear detonations, have wavelengths smaller than X-rays.

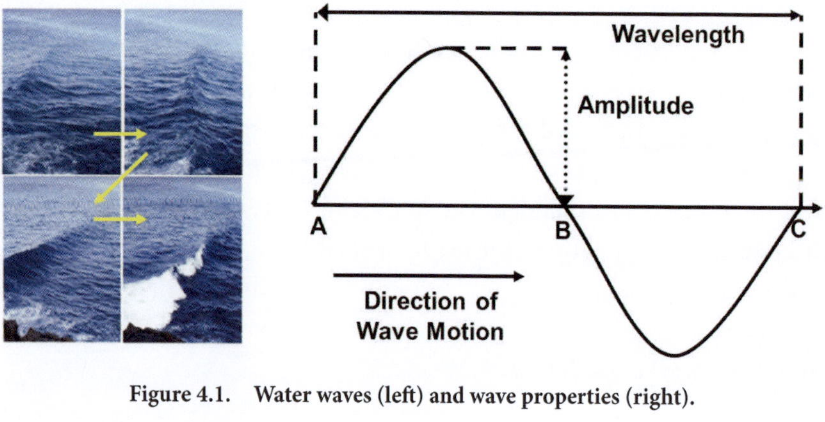

Figure 4.1. Water waves (left) and wave properties (right).

This wide range of wavelengths comprises what is known as the electromagnetic spectrum.

The speed of light for all wavelengths of the electromagnetic spectrum is the product of the wavelength and frequency. For example, different colors of visible light have different wavelengths and different frequencies. The frequency of a light wave is proportional to its energy.

The cosmic microwave background (CMB) radiation we encountered in Chapter 2 appears to be black body radiation. An example of a black body is shown in Figure 4.2. It is an insulated, hollow enclosure, or cavity, with a small hole in one side. The exterior walls of the cavity do not allow heat or radiation to enter the cavity. The interior walls must be able to absorb all radiation frequencies incident upon them and radiate energy that is characteristic of the temperature of the cavity. We study black body radiation by measuring the wavelength of the radiation which escapes from the small hole in the side of the cavity.

Energy density of the emitted radiation is plotted against wavelength in Figure 4.2. The curve labeled "UV Catastrophe" in the figure shows the calculation of the energy density of black body radiation based on the classical theory of physics. Theory based on classical physics assumes that energy is a continuous function of wavelength. The classical theory predicts an extremely large energy density of black body radiation emitted at very short wavelengths. Instead, experiments show the energy density becomes small as the wavelength approaches zero. The discrepancy between classical theory and experimental observations at low wavelengths is known as the "ultraviolet catastrophe." Is something wrong with the classical theory?

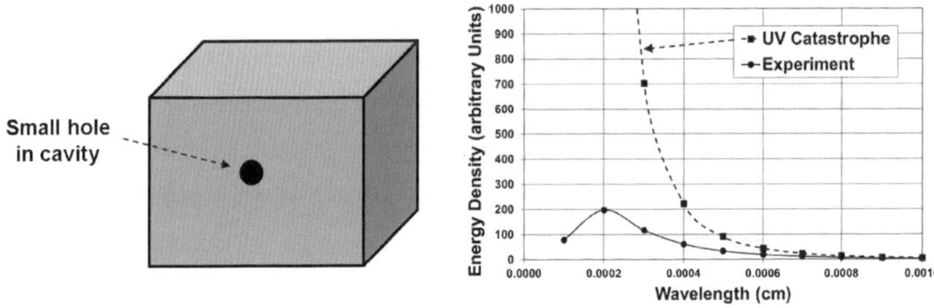

Figure 4.2. Black body radiation.

German physicist Max Planck (1858–1947) provided an answer to this question in 1900. The curve labeled "Experiment" in Figure 4.2 shows a theoretical match of experimental data, including observations of emitted radiation for very small wavelengths. To achieve the theoretical match, Planck was compelled to replace the assumption that energy is a continuous function of wavelength with a new and somewhat bold concept: radiation exists in discrete packets of energy. These discrete energy packets are called quanta and the theory of which quanta are a part is called quantum theory.[3] Planck's hypothesis that a quantum of light is a discrete packet of energy has been substantiated by many experimental tests and applications.

The Uncertainty Principle

One of the most peculiar and philosophically significant consequences of the quantum theory is the uncertainty principle proposed by German physicist Werner Heisenberg (1901–1976)[2] in 1927. Heisenberg [1927] showed that simultaneous measurements of two complementary physical variables would have a characteristic uncertainty. The uncertainty would limit how precisely we can measure values for such pairs of variables as position and momentum or energy and time.

The energy-time uncertainty relation says that there is a limit to how well we can simultaneously measure the energy in a quantum process in a measured duration of time. Another way to understand the energy-time uncertainty relation is to recognize that a large amount of energy can appear for a short duration of time specified by the energy-time uncertainty relation. Particles that appear for the short duration allowed by the energy-time uncertainty relation are virtual particles.

The energy-time uncertainty relation is needed to understand the vacuum process in Figure 3.5. The vacuum process shows the creation, interaction, and annihilation of a proton (p), an antineutron ($\overline{n}$), and a pion (π). No particles exist before or after the vacuum process shown in the figure. The particles in the vacuum process can exist for a short duration of time as long as the energy-time uncertainty relation is satisfied. This appears to be true even if the conservation of energy is violated for the duration allowed by the energy-time uncertainty relation. The short duration, random change in energy at a point in space is an example of a quantum fluctuation. Quantum fluctuations can serve as a natural mechanism for initiating the Big Bang, as we discuss further in Chapter 5.

Heisenberg suggested a limit exists to the level of knowledge we can obtain about nature. This limitation is not something we can reduce by improving our measuring techniques. It is a limitation imposed by nature. On the scale of submicroscopic particles, such as the electron and the photon, it is not possible to arbitrarily reduce the effect of the experimentalist on the system being observed. As Heisenberg noted:

> "... in classical physics it has always been assumed either that this interaction [between observer and observed object] is negligibly small, or else that its effect can be eliminated from the result by calculations based on 'control' experiments. This assumption is not permissible in atomic physics; the interaction between observer and object causes uncontrollable and large changes in the system being observed, because of the discontinuous changes characteristic of atomic processes." [Sambursky, 1975, p. 518]

Scientists now believe that uncertainty relations apply to all phenomena.

Classical physics has successfully described a wide range of natural phenomena, particularly macroscopic phenomena. Quantum theory accounts for the success of classical physics by containing the classical theory as a special case of the more comprehensive quantum theory.

Point to Ponder: Does the Uncertainty Principle Have a Philosophical Impact?

French physicist and Prince Louis Victor de Broglie (1785–1870) observed that Heisenberg's uncertainty principle makes it "impossible to attribute simultaneously to a body a well-defined motion and a well-determined place in space and time" [de Broglie, 1966, p. 122]. If we cannot specify with certainty both the location and the motion of an object at a given time, then the determinism of classical physics fails. The failure of determinism in classical physics cast doubt on mechanistic or deterministic philosophies that relied on classical physics.

The Particle Zoo

The processes diagrammed in Figure 3.5 have introduced six particles: the electron, positron, photon, proton, pion, and antineutron. These particles are but six residents of the particle zoo.

Scattering experiments have led to the discovery of a large number of particles.[1] Most of these particles are unstable because they do not exist very long. Unlike virtual particles, the lifetimes of unstable particles are not determined by the uncertainty principle. The interaction associated with the decay of unstable particles is the weak interaction. It is one of four fundamental interactions (Figure 4.3). In order of decreasing strength, the four recognized interactions are the strong interaction, the electromagnetic interaction, the weak interaction, and the gravitational interaction. The electromagnetic and weak interactions are often combined and called the electroweak interaction.

The term "interaction" stems from the theoretical interpretation of a force between two objects as an exchange of particles. For example, the Earth and the Moon are thought to exchange particles called gravitons. The physical consequence of the exchange of gravitons is an attraction between the Earth and the Moon. Examples of the fundamental interactions are presented in Figure 4.3.

The strong interaction is the interaction that binds nuclei together. All atoms consist of a nucleus, which contains most of the mass of the atom,

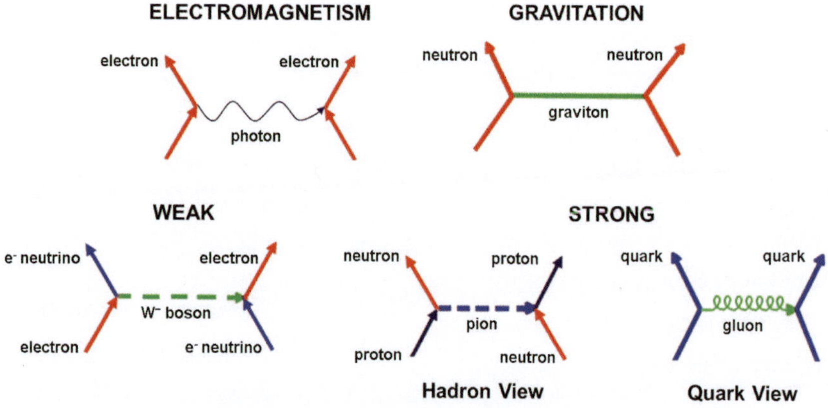

Figure 4.3. Fundamental interactions.

and an electron cloud surrounding the nucleus. Positively charged protons and particles without an electrical charge — neutrons — are the principal building blocks within the nucleus. The electron, proton, and neutron were the only particles needed to populate the Periodic Table of elements that was first formulated by Russian chemist Dmitri Mendeleev (1834–1907) in the latter half of the 1800s.

Negatively charged electrons occupy regions of space surrounding the nucleus. The actual location of the electrons is not known because of Heisenberg's uncertainty principle. Instead, we calculate the probability of finding an electron at a particular spatial location. The term "electron cloud" is used to denote both our ignorance of exact spatial locations and our knowledge of probability distributions. The volume of space occupied by the nucleus is very small in comparison to the volume of space occupied by the electron cloud. The electromagnetic attraction between protons and electrons is primarily responsible for maintaining the electron cloud around the nucleus. By contrast, electromagnetic repulsion between protons in the small volume of the nucleus would disrupt the nucleus if another force was not present. The strong interaction is that other force. Both protons and neutrons are carriers of the strong force. Nuclear power, both peaceful and military, depends on disrupting the strong force-binding particles in the nucleus of an atom.

The strong and weak interactions are short-range forces. Their influence is dominant only when the interacting particles are very close to each other: at distances comparable to the diameter of a nucleus or during the first few moments after the Big Bang. The electromagnetic and gravitational interactions are long-range forces. They exert an influence on particles separated by distances ranging from the diameter of an atom to the diameter of the universe. The many known particles can be classified by their interactions.

Table 4.1 presents a classification scheme based on strong and weak interactions that can help us understand the particle zoo. Level 1 shows two general categories of particles with mass. Massless particles of light (photons) are in a category of their own. Massive particles in Level 1 are called either leptons or hadrons. Hadrons can participate in the strong interaction; leptons cannot. Examples of leptons are the neutrino, the electron, and the positron. Neutrinos do not interact electromagnetically. Electrons and positrons have electrical charges and interact electromagnetically.

Table 4.1 Classification of particles based on strong and weak interactions.

Level 1	Level 2	Level 3	Level 4
Photons γ			
Leptons (No strong interaction)	Neutrino ν		
	Electron e^- Positron e^+		
Hadrons (Strongly interact)	Mesons (Interact weakly)	Pion π	
		Kaon K	
		Eta η	
	Baryons (Distinct particle–antiparticle pairs)	Nucleon	Proton p
			Neutron n
		Lambda Λ Sigma Σ	

The two hadron subgroups in Level 2 are mesons and baryons. Mesons can interact weakly, while baryons do not. Mesons, such as the pion, can be their own antiparticle, whereas baryons exist in distinct particle–antiparticle pairs. An example of a baryon pair is the proton-antiproton. Another common baryon is the neutron. The proton and neutron are also called nucleons because of their prominent role in atomic nuclei. The proton is a good illustration of the classification levels in Table 4.1. Starting with the proton in Level 4, we see that the proton is a nucleon; nucleons are baryons; and baryons are hadrons.

The gravitational and electromagnetic interactions are not as well suited for classifying particles as the strong and weak interactions. For example, all particles with mass can participate in gravitational interactions. By contrast, only electrically charged particles, such as electrons and protons, participate in electromagnetic interactions.

Atomism and Quarks

Historically the treatment of matter as an aggregate of tiny, indivisible particles has been a useful approach. The conceptual usefulness and proven success of the notion that elementary, indivisible particles exist is often grasped by people searching for an understanding of unexplained phenomena. As technology progressed, objects once thought to be elementary are found to be composed of even more fundamental objects. Hadrons are an example.

Prior to the beginning of the twentieth century, atoms seemed to be the fundamental particles of nature. By the turn of the century, a number of observations had been made which could best be explained if atoms were thought of as objects comprised of smaller objects. The electron, proton, and neutron were born.

The electron, proton, and neutron were sufficient for explaining the results of most experiments for decades. One observation in need of an explanation was the existence of stable nuclei. A stable nucleus is a nucleus that does not decay into smaller fragments if it is left alone. Nuclear stability was eventually explained in 1935 when Japanese physicist Hideki Yukawa (1907–1981) postulated the existence of a new particle called the pi meson.[1,4] In 1936, a particle having many of the properties of Yukawa's pi meson was discovered. Further analysis of the properties of this particle showed it did not interact strongly with other matter. Yukawa's pi meson was not actually discovered until 1947 when it was observed as a constituent of cosmic radiation by British physicist Cecil F. Powell (1903–1969). By then people realized that the number of elementary particles was much greater than originally thought.

The existence of so many particles prompted people to look for particles of a more elementary nature. The quark model, conceived independently by Murray Gell-Mann (1929–2019) and George Zweig (1937–), appeared in 1964. Gell-Mann borrowed the word "quark" from James Joyce's (1882–1941) *Finnegan's Wake* which was published in 1939.

According to the quark model, mesons such as the pi meson, or pion for short, are composed of a quark and an antiquark. Baryons such as the proton and neutron are composed of three quarks. By postulating the existence of a relatively small number of quarks, it was possible in the late 1960s and 1970s to construct the many hadrons that have been observed. Thus, the number of "truly elementary" particles was reduced to the number of quarks and leptons. Today the number of quarks and leptons has grown from the original two quarks and two leptons to six "flavors" of quarks and six kinds of leptons.

Figure 4.4 illustrates the constituents of an atom. The atom consists of a nucleus and electrons. The nucleus contains the strongly interacting nucleons, namely protons and neutrons. The quark model says that a proton contains two up quarks and a down quark, while a neutron contains one up quark and two down quarks. The table in the figure shows that everyday matter is composed of up and down quarks, and two leptons. The leptons are electrons and electron-neutrinos (e-neutrino). More exotic matter produced in high energy processes is composed of four quarks and four leptons. The electron, muon, and tau leptons are electrically charged, while neutrinos are electrically neutral.

An important property of quarks is the postulated value of their electrical charge. The charge of every observed particle with an electric charge has the same magnitude as the charge of the electron. This experimental fact is the basis for the historical practice of calling the electrical charge of the electron the elementary charge. The electrical charge of quarks is thought to be a fraction of the elementary charge of the electron. The difference in magnitude of the electrical charge provides a means for searching for and identifying quarks. To date, indisputable experimental proof of the observation of a particle with an electrical charge less than the elementary charge is not available.

The evidence for quarks is much more comprehensive than just a search for fractional elementary charge. Highly energetic scattering experiments have shown that nucleons — protons and neutrons — exhibit an internal structure. The internal structure can be understood if we think of nucleons as composites of point particles: quarks.

Hadrons in Table 4.1 are made up of quarks. A meson is a composite of a quark and an antiquark, and a baryon is a composite of three quarks. Quarks seem to be bound in pairs and triplets. Attempts to produce unbound quarks have been made at particle accelerators around the world. Such attempts to

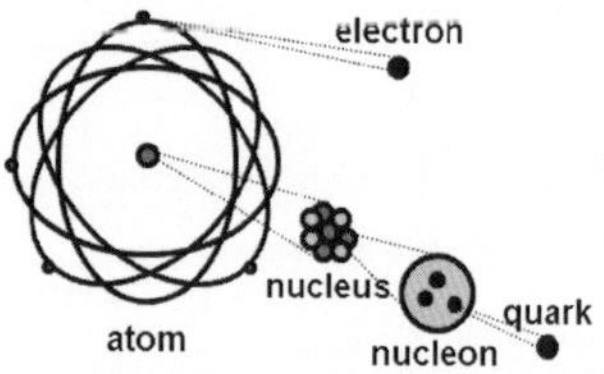

	QUARKS	LEPTONS
Everyday Matter	Up, down	Electron, e-neutrino
High Energy Matter	Strange, charm Bottom, top	Muon, mu-neutrino Tau, tau-neutrino

Figure 4.4. Constituents of matter.

produce unbound quarks have been unsuccessful. If quarks exist, as seems necessary based on a wealth of experimental data, why have unbound, or free, quarks not been observed?

The most plausible explanation for the failure to observe free quarks is a mechanism called asymptotic freedom. Simply stated, asymptotic freedom is the idea that the strength of the interaction between quarks increases as the quarks move away from each other. Conversely, the interaction between quarks decreases as bound quarks approach each other. If the asymptotic freedom concept is correct, experiments capable of scattering a particle from the quark system should show quarks behaving like free particles within the region of space they occupy. This type of behavior is analogous to the behavior of two balls bound together by a rubber band. If the balls are near each other, the rubber band is slack and the balls behave as if they are free. If we pull the balls apart, the rubber band would resist their separation. If the balls are released, they would be pulled together again. The only way to completely separate the balls would be to break the rubber band. A similar requirement is suggested for quarks: to separate quarks, it is necessary to disrupt the strong force binding them.

To date, experiments have not split a hadron into its quark constituents, although they have shown that quarks behave like free particles inside the hadron. The quark model is supported by considerable experimental evidence, but no free quarks have been observed. A proof that free quarks and antiquarks do not exist because they have all combined to form hadrons depends on the origin of quarks and antiquarks, which depends in turn on the origin of the universe.

Point to Ponder: Are Quarks Elementary Particles?

The possible existence of other levels of structure more fundamental than quarks and leptons has been proposed. Here we consider only quarks and leptons because they are part of the standard model of particle physics. We should keep in mind that quarks may not be the last step along the path toward finding truly elementary and indivisible particles. In fact, when we discuss the standard model of particle physics, we shall see that quark fields are considered more fundamental than quarks as particles.

Standard Model of Particle Physics

Quantum mechanics and special relativity were two of the great achievements of physics in the first half of the twentieth century. Quantum mechanics explained the behavior of submicroscopic systems, and special relativity described the behavior of objects moving at speeds approaching the speed of light. The two theories were first combined in the 1920s to form a theory now known as relativistic quantum theory.[5]

Relativistic Quantum Theory

Relativistic quantum theory had to reconcile seemingly incompatible observations. For example, the number of particles is conserved in a nonrelativistic quantum mechanical system. By contrast, energy can be converted into mass and mass converted into energy in relativistic systems. These observations showed that some previously accepted conservation laws were no longer true.

A conservation law can be surmised if a measurable physical property in a closed system does not change with time. An example of a conservation law is conservation of electric charge. Conservation laws can change as new observations are made. Conservation of energy, for example, says that the energy of a closed system does not change with time. It was considered a valid conservation law until scientists observed that the number and type of particles can change when particles are created or destroyed. The correct conserved property is mass-energy, that is, the combination of mass and energy. The identification of conservation laws is an important task of contemporary high energy physics.

Relativistic quantum theory accounts for mass-energy conversion. The most widely accepted relativistic quantum theory is the standard model of particle physics, which for brevity is called the Standard Model. The quark model is now a component of the Standard Model.

The Standard Model

The Particle Data Group (PDG) [Zyla *et al.*, 2020] periodically compiles experimental results and a selection of theoretical attempts to understand the results. The PDG includes precision tests of quantum electrodynamics (QED), which is a theory of the interaction of electrons and photons. Precision tests of QED show that the theory agrees with measurements to an accuracy of one part per billion or better. QED is one of the most accurate physical theories ever developed for studying scattering systems. It is an example of a quantum field theory (QFT) and is part of the Standard Model. It is also worth noting that the mathematical equations of QFT are difficult to solve when applied to bound state systems such as the behavior of electrons bound to the nucleus of an atom.

So far, we have discussed particles as if they were real, physically indivisible objects. The modern view is that particles are actually quantized excitations of fields. A field has a value at every point in spacetime, even if that value is zero.

There is a conceptual conflict between the concepts of particles and fields. If we consider the example of an electron interacting with a photon, our goal is to treat the electron and the photon on an equal footing: they are either both particles or both fields. If we assume that particles are fundamental, then the electromagnetic field arises from a collection of photons. On the other hand, if we assume that fields are fundamental, then each photon arises from quantization of the electromagnetic field, and each electron arises from the quantization of a matter field.

The Standard Model is a QFT of strong, and electroweak interactions. It assumes the existence of quarks, leptons, and the Higgs particle (Table 4.2). The Higgs particle is named after British physicist Peter Higgs (1929–), who pioneered its use. The fermions (quarks and leptons) are not allowed

Table 4.2 Fundamental particles of the standard model.

Fermions (matter particles)				**Gauge Bosons (force carriers)**	**Higgs Boson (origin of mass)**
	Generation				
	1	2	3		
Quarks	u	c	t	Photon γ	Higgs H
	d	s	b	Gluon g	
Leptons	ν_e	ν_μ	ν_τ	Z^0 boson	
	e	μ	τ	$W^\pm$ boson	

to simultaneously occupy the same quantum state, while bosons can simultaneously occupy the same quantum state. Matter particles acquire mass by interacting with the Higgs boson, which was discovered in 2012 [CMS Collaboration, 2012]. Interactions between fermions are mediated by force carriers. For example, the photon mediates the electromagnetic interaction, the gluon mediates the strong interaction, and the weak interaction is mediated by the $W^{\pm}$ and Z^0 bosons. Figure 4.3 illustrates some of these interactions.

According to QFT, each of the "particles" shown in Table 4.2 should be viewed as an excitation of a corresponding field, and the "force carriers" are excitations that mediate interactions. For example, the electron is an excitation of the electron field and the Higgs boson is an excitation of the Higgs boson field. A particle interaction occurs when there is a coupling between fields. Particle mass is a measure of the energy needed to make the field vibrate. A more massive field can decay into less massive fields. Two of these concepts are illustrated in Figure 4.5.

We observe particles even though the reality of QFT consists of quantum fields. Figure 4.6 shows that the Higgs field is different from all other fields because the resting value of the Higgs field in empty space is nonzero. The resting value is also known as the vacuum expectation value. The resting value of the other fields is zero. All of the fields except the Higgs field would move at the speed of light if the Higgs field was not present. In the presence of the Higgs field, the other fields, except for the photon and possibly the neutrino, acquire mass by coupling to the Higgs field. A massive neutrino is not part of the Standard Model. A Higgs boson is shown as an excitation of the Higgs field in Figure 4.6.

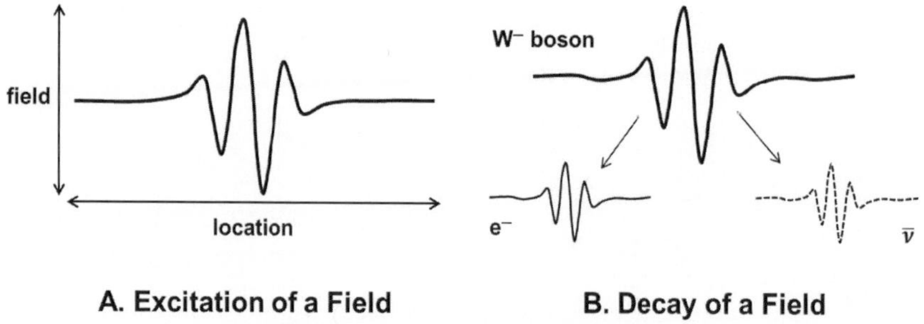

Figure 4.5. Excitation and decay of a field.

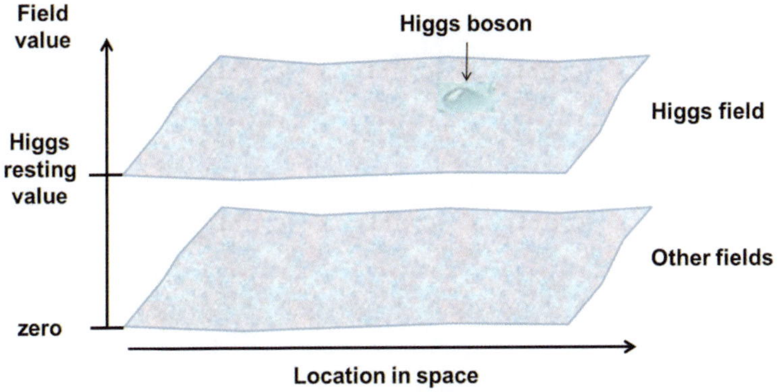

Figure 4.6. The Higgs field [modified from Carroll, 2012, p. 34].

Despite its record of success, many anomalies can be identified in the Standard Model. A few examples are listed below:

- Masses of quarks, leptons, and the Higgs particle are some of the parameters that are not predicted by the theory and must be entered into the Standard Model.
- The Standard Model assumes that neutrinos are massless, yet experiments imply that neutrinos have mass. Neutrinos with mass can be included in an extension of the Standard Model, but this requires the use of more unexplained parameters.
- The Standard Model does not account for gravity. A theory of gravity is needed that is consistent with QFT and is able to yield general relativity as the classical approximation. The graviton shown in Figure 4.3 has not yet been observed.
- The Standard Model does not explicitly account for dark matter and dark energy.

Proponents of the Standard Model recognize that anomalies may change the mainstream paradigm.

The Big Bang — The Moment of Creation

We saw in Chapter 2 that Georges Lemaître (1894–1966) developed a model universe from Einstein's general relativistic equations. Lemaître's model could represent a static, expanding, or contracting universe. The interpretation of Hubble's law seemed to favor the idea of a universe that was expanding from a much more compact accumulation of mass and energy. The hypothesis of the primeval atom by Lemaître in 1931 provided a mechanism for initiating an explosive expansion of the universe that began in a moment of creation. Observations of recessional velocity-distance relations and the CMB further supported the idea of a Big Bang.

There are many unproven hypotheses about what happened at the moment of the Big Bang. One hypothesis is that all of the mass-energy of the universe was initially a hot, dense state located at a point in spacetime called a singularity. This idea is similar to Lemaître's primeval atom. The source of the initial mass-energy is not known. Several nonsecular suggestions are considered in later chapters. Here we focus on the possibility that the entire universe is the outgrowth of a quantum process.

Quantum fluctuations are a natural mechanism for explaining the origin of the universe.[6] A quantum fluctuation is a short duration, random change in energy at a point in space. The duration and change in energy are constrained by the energy-time uncertainty relation. Quantum fluctuations in the vacuum, such as a vacuum process, can distort spacetime and produce a locally dense accumulation of particles known as a quantum foam. The quantum foam consists of numerous quantum fluctuations in a region of space and duration of time that are constrained by uncertainty relations. If a fluctuation is self-sustaining, the presence of matter can impart a local expansion of spacetime, which creates more room for additional fluctuations in a snow-balling expansion of the universe. According to this view, the quantum foam initiated the Big Bang.

❁ ❁ ❁

Finding an explanation for the number and type of particles that have been observed and the many ways they interact is a challenging and ongoing problem. The most widely accepted theory of particle physics is the Standard Model. More exotic theories, such as string theory, M-theory, supersymmetry, supergravity, and grand unified theories, have been proposed.[7] None of the theories has been indisputably established. As long as we are intellectually honest enough to recognize the limitations of the theories we use, we can maintain a proper perspective in our quest to understand the origin and evolution of the physical universe.

Point to Ponder: Science or Science Fiction?

Some physicists believe that some of the leading interpretations of modern physics are so far removed from our traditional understanding of nature that they are more like science fiction than science.[8] For example, Richter [2006, p. 8], observed that "much of what currently passes as the most advanced theory looks to be more theological speculation." In *Farewell to Reality*, Baggott [2013, p. 2] presented what he called the authorized version of reality in Part I, and introduces the idea of fairy tale physics in Part II where he discusses exotic theories such as those mentioned above. Baggott argued that fairy tale theories tend to be "more metaphysics than physics" [Baggott, 2013, p. 205], and concluded "that in fairy tale physics the metaphysics is all there is" [*Ibid.*, p. 287]. Baggott mused in the preface: "If we were to regard fairy tale physics as a lively branch of contemporary philosophy rather than science, do you think it would continue to receive the same level of attention from funding agencies, universities, popular science publishers, the producers of radio and television programs, and the wider program? No?"

Chapter 5

The Inflation Model of Cosmology

The most widely accepted scientific model of the evolution of the universe is the inflation model of cosmology.[1] It is presented in this chapter using the evidence and theories discussed in previous chapters. We used the standard model of particle physics and vacuum fluctuations to initiate the Big Bang in Chapter 4. The scientific view of the evolution of the universe depends on cosmological models. Following a discussion of two leading cosmological models, we summarize the cosmological history of the physical universe.

The Standard Model of Cosmology

Many models have been proposed as models of the aftermath of the Big Bang. In this section, we consider two models that have made a significant impact on our understanding of cosmology: the standard model of cosmology, and the Lambda-CDM (ΛCDM) model. According to the standard model of cosmology summarized in Table 5.1, the universe we see today began with the Big Bang and underwent a uniform expansion of spacetime. The model assumes that ordinary matter was the only known type of matter, and the universe obeys the cosmological principle, that is, the spatial distribution of matter is homogeneous and isotropic.

Table 5.1 The standard model of cosmology.

Event	Comment
Big Bang	Origin of cosmic microwave background
Hubble expansion	Uniform expansion of spacetime Ordinary matter is only type of matter

The standard model of cosmology was able to account for universal expansion, but other issues remained, and pointed to the need for a revised cosmological model. We consider a few of the issues that motivated revisions to the standard model of cosmology.

Issues in Cosmology

An expanding universe model explains the redshift observed by astronomers, but it raises new issues. For example, what caused the expansion? A more subtle issue is the implication that the universe may be curved and finite. Cosmological models based on Albert Einstein's (1879–1955) general theory of relativity treat the universe as both curved and finite in four-dimensional spacetime. Figure 5.1 presents possible geometries of the universe and an illustration of the distribution of temperature fluctuations associated with each geometry. The flat geometry is Euclidean geometry, while open and closed geometries are non-Euclidean geometries.

The observed density of matter in the universe implies that the universe has a flat geometry except near concentrations of matter such as stars and planets. Dark matter adds to the observed matter density of the universe and helps explain why the critical density of the universe is one, which is the critical density of a flat universe. Dark energy, on the other hand, helps explain why the expansion of the universe is accelerating. When viewed

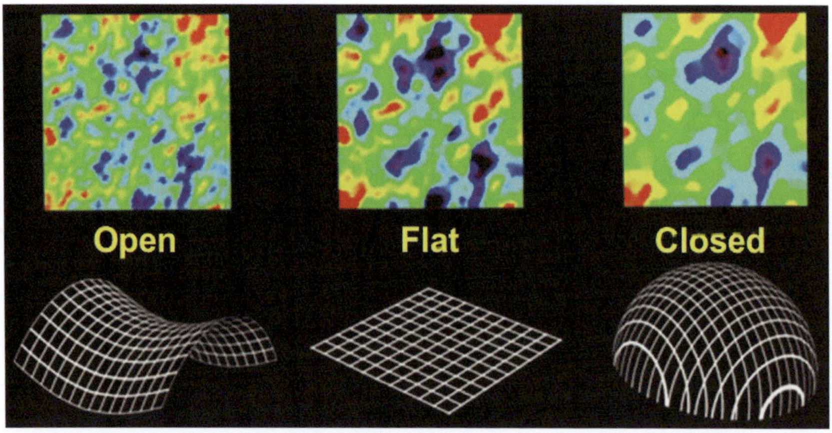

Figure 5.1. Possible geometries of the universe [NASA-WMAP, 2010].

within the context of general relativity, dark energy appears to behave as a gravitationally repulsive form of energy that exerts a negative pressure.

Dark matter and dark energy can be introduced into the cosmological model using Einstein's cosmological constant. The cosmological constant is zero in the standard model of cosmology. It is now believed that a nonzero value of the cosmological constant is needed to account for dark matter and dark energy.

The horizon problem is another problem with the standard model of cosmology. The horizon problem is based on the recognition that some regions in the universe are spacelike separated. These regions are so far from one another that they cannot communicate at the fastest possible speed, which is the speed of light in vacuum. We could send a light single from one of these regions to another and our signal would never be received because the separation between our region and the target region is large and continuing to grow as the universe expands. We see only the observable universe; everything else is beyond our horizon. Given the horizon problem, how can the universe be in nearly ideal thermodynamic equilibrium, which is implied by the homogeneity and isotropy of the cosmic microwave background (CMB) radiation? A period of rapid inflation in the early universe provides a period when all of the universe was close enough to share information.

A revised model of cosmology that can resolve such issues as the flatness problem and the horizon problem should include inflation and a nonzero cosmological constant. The ΛCDM model presented next satisfies these requirements.

The Lambda-CDM Cosmological Model

The scientific worldview incorporates observations of the cosmological redshift, the CMB, and the distribution of matter in the universe. Measurements of anisotropy in the cosmic background radiation (CBR) using the Wilkinson Microwave Anisotropy Probe (WMAP) have provided more data for developing improved cosmological models. By combining the WMAP measurements with cosmological models, cosmologists are finding evidence to support the view that we live in a flat universe that underwent an inflationary expansion period immediately after the Big Bang. A flat universe occurs when the average density of the universe equals the critical density. Furthermore, the expansion of the universe may be accelerating. This view, however, depends on a poorly understood type of energy known as dark energy.

The inflation cosmological model known as the ΛCDM cosmological model inserts an inflationary period of rapid spacetime expansion shortly after the Big Bang and before the uniform spacetime expansion period associated with the Hubble expansion rate. Lambda denotes the cosmological constant, which represents the unknown cause of inflationary expansion. The unknown cause is called dark energy and accounts for approximately 70% of the energy density of the universe. CDM stands for cold, dark matter and accounts for approximately 25% of the energy density of the universe. The remaining 5% of the energy density of the universe is the mass and energy that we observe as ordinary matter.

The ΛCDM model is summarized in Table 5.2 and Figure 5.2. The dark matter halo referred to in the table is an accumulation of dark matter.

Table 5.2 The Lambda-CDM cosmological model.

Event	Comment
Big Bang	Origin of cosmic background radiation
Inflation	Rapid expansion of spacetime caused by dark energy (nonzero cosmological constant)
Hubble expansion	Uniform expansion of spacetime Formation of dark matter halos Formation of ordinary matter objects, such as nebulae, galaxies, and star clusters

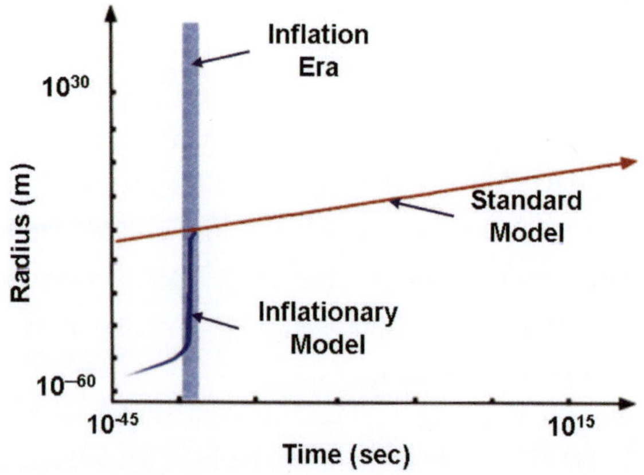

Figure 5.2. Inflationary expansion of the universe [Guth, 2002].

Ordinary matter is present in the dark matter halo as a hot, gaseous phase. Within the context of the ΛCDM model, ordinary matter stars and galaxies form inside the dark matter halo as the gaseous phase of ordinary matter cools.

Figure 5.2 illustrates a comparison of the inflationary model with the standard model. The inflationary model begins with a smaller radius than the standard model and rapidly expands during an inflation era until it matches the size of the standard model. The mechanism for inflation is not yet known.

Figure 5.3 illustrates several features of Big Bang expansion. For example, the WMAP spacecraft is shown on the right-hand side of the figure in the present. It is observing light emanating from the afterglow light pattern shown at approximately 375,000 years after the Big Bang, which is depicted in the figure as quantum fluctuations. The quantum fluctuations are immediately followed by an inflationary period prior to the afterglow light pattern. Other features in the figure are discussed below as part of the cosmological history of the physical universe. First, we must consider the behavior of the four fundamental interactions as the universe expands and cools. Theories

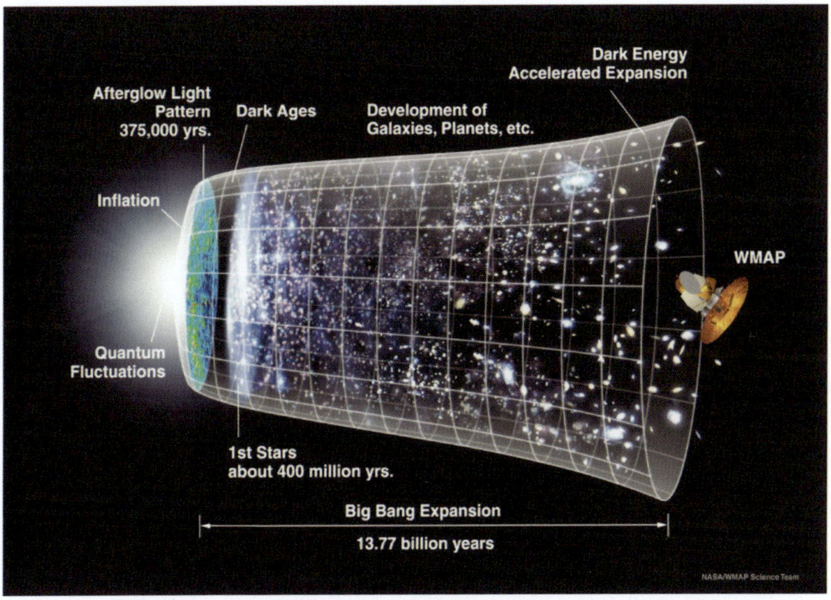

Figure 5.3. The Lambda-CDM model [NASA-WMAP, 2010].

known as the Grand Unified Theory (GUT) and Theory of Everything (TOE) explain how the four fundamental interactions emerge over cosmic time from a unified interaction. These theories help us understand different phases associated with the inflation model of cosmology.

Grand Unified Theory and a Theory of Everything

We saw in Chapter 4 that four fundamental interactions have been identified: strong, electromagnetic (EM), weak, and gravitational. The mainstream view of physics is that a theory can be formulated which will unify these four seemingly disparate interactions into a single type of interaction.[2] Many people have tried and continue to try to formulate such a law. Einstein, for example, attempted to unify the gravitational and EM interactions in his formulation of General Relativity. He failed. No one has succeeded in unifying all four interactions. The status of unification of interactions is considered here.

Steven Weinberg (1933–), Abdus Salam (1926–1996), and Sheldon Lee Glashow (1932–) were awarded the 1979 Nobel Prize in Physics for their work in developing a unified theory of the weak and EM interactions. The combination of interactions is now known as the electroweak interaction and spurred efforts to unify the remaining two interactions. Theories that attempt to combine strong, EM, and weak interactions are called Grand Unified Theories. Glashow challenged the term when he observed that the gravitational interaction remains elusive:

> "... a synthesis of weak, strong, and electromagnetic interactions has been called a 'grand unified theory', but a theory is neither grand nor unified unless it includes a description of gravitational phenomena. We are still far from Einstein's truly grand design." [Glashow, 1980, p. 539]

A theory that combines all four interactions is dubbed a Theory of Everything. TOE should include a GUT and a theory of gravity as special cases, as illustrated in Figure 5.4. The GUT combines the strong and electroweak interactions. The strong interaction is described by a quantum field theory known as quantum chromodynamics (QCD) that uses gluons to mediate the interaction between quarks. The theory associates the coupling of the strong interaction with a quantum number called the color charge that is assigned to quarks and gluons. The electroweak interaction combines the EM and weak interactions. Quantum Electrodynamics (QED) is the quantum field theory of the EM interaction.

Figure 5.4 shows the dominant interactions in the universe at different points in time along a time axis that ranges from the moment of the "Big Bang" to the present. The dominant interactions and corresponding theories change as the universe cools. Cooling occurs as a result of the evolution and expansion of the universe. The decrease in temperature associated with cooling allows transitions to occur from one interaction to other interactions. In the figure, TOE transitions to the gravitational interaction and GUT, which then transitions into the strong and electroweak interaction. As the universe continues to cool, the electroweak interaction transitions into the EM and weak interactions.

It is now thought that the transitions shown in Figure 5.4 occur when there is a change in symmetry of a quantum state. In this context, symmetry refers to a state that is invariant with respect to a set of transformations, such as the Lorentz transformation of spacetime. Physicist Philip Anderson pointed out that the state "must always have the same symmetry as the laws of motion which govern it" [Anderson, 1972, p. 394]. In our case, the universe enters a new phase when a transition from one dominant interaction to another dominant interaction occurs. The symmetry of the initial state changes as a result of the transition and the change is known as symmetry breaking.

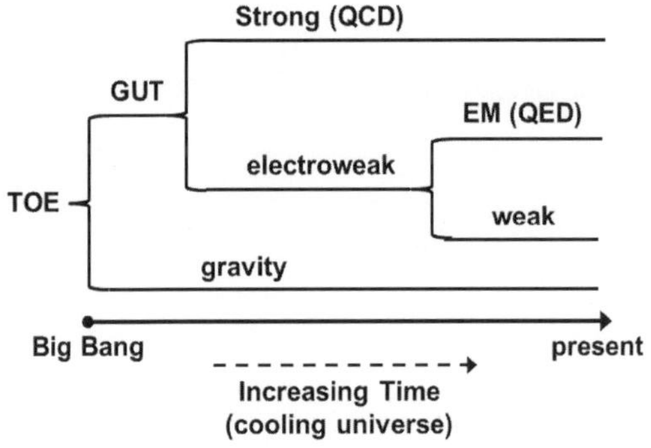

Figure 5.4. Theory of everything.

One of the currently promising approaches to a TOE is quantum gravity,[3] which encompasses theories that are attempting to restructure general relativity according to principles of quantum theory. The search for a TOE continues.

The Radiation-Dominated Era

A cosmological history of the physical universe can be organized in terms of a cosmic timeline. Cosmic time, also known as cosmological time, can be thought of as the time measured by clocks that move with a homogeneous, expanding universe. Our reckoning of cosmic time begins with the Big Bang.

Cosmological history can be subdivided into three eras that depend on the mass density of radiation, matter, and vacuum energy as illustrated in Figure 5.5. The mass density of radiation consists of the mass equivalent of photon energy produced by the Big Bang. The mass density of matter consists of the mass densities of ordinary matter and dark matter. The mass density of vacuum energy refers to the mass equivalent of vacuum energy associated

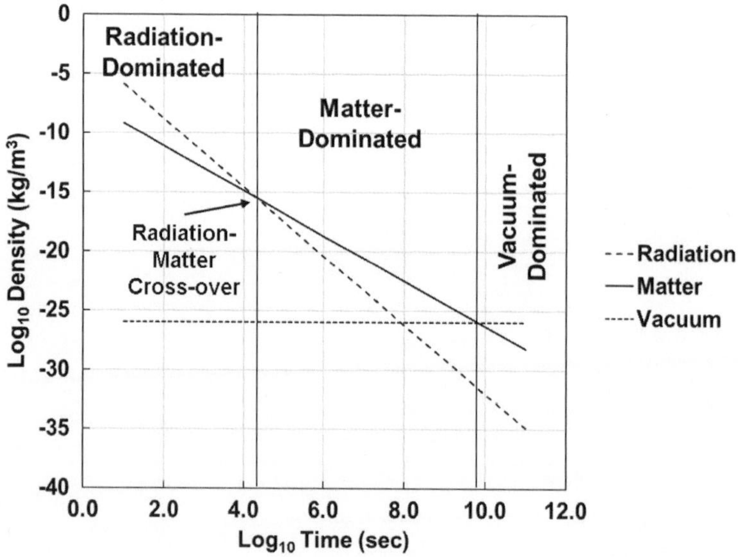

Figure 5.5. Competition between radiation, matter, and vacuum eras [Perlov and Vilenkin, 2017, p. 170].

with dark energy and quantified by a nonzero cosmological constant. The values in Figure 5.5 are subject to revision as research progresses. In this section, we summarize the chronology of the Radiation-Dominated Era.

Radiation-Dominated Era

Primordial energy, the initial mass-energy of the universe, occupied a singularity, which is an infinitesimally small volume of space. The expansion of the universe after the Big Bang propelled primordial energy outward from the singularity. As cosmic time increased, the size of the universe increased, the distance between particles increased, and the temperature of the universe decreased.

Mass-energy includes radiation, ordinary matter, dark matter, and dark energy. The mass density of radiation is the largest mass density in the Radiation-Dominated Era. Figure 5.5 shows that the transition from one era to another depends on which mass density is largest. Radiation from this era includes the Cosmic Background Radiation. Matter is embedded within the CBR during the Radiation-Dominated Era. The pressure exerted by radiation exceeds the pressure exerted by matter or vacuum energy. The CBR has cooled and exists today as the CMB.

Table 5.3 presents key events in the Radiation-Dominated Era. Cosmic times in the table are approximate and subject to revision. Keep in mind that the increase in cosmic time is accompanied by universal expansion and cooling of the temperature of the universe.

Matter–Antimatter Asymmetry

The preponderance of matter in the observed universe implies that there must have been an imbalance between matter and antimatter. This is referred to as the matter–antimatter asymmetry in Table 5.3 and is worth special attention. Primordial energy in the early moments following the Big Bang existed primarily in the form of EM radiation. We saw in Chapter 4 that matter-antimatter pairs can be produced from EM energy and annihilated to produce EM energy. The ratio of matter to antimatter was expected to be one to one based on the pair production and annihilation processes.

Table 5.3 **Radiation-Dominated Era.**

Epoch	Cosmic Time (end of Epoch)	Comment(s)
Big Bang	0 sec	Possibly initiated by quantum fluctuations
Planck	10^{-43} sec	Origin of space, time, and energy. The laws of physics may not apply
GUT	10^{-35} sec	TOE separates into gravity and GUT interactions
Inflation	10^{-33} sec	Rapid expansion of spacetime possibly caused by dark energy (nonzero cosmological constant)
Electroweak	10^{-10} sec	GUT separates into strong and electroweak interactions
Quark	10^{-5} sec	The electroweak interaction separates into weak and electromagnetic interactions. The universe contains a soup of quarks, leptons, photons, and other particles
Hadron	1 sec	Quarks are bound by the strong interaction and form hadrons such as protons and neutrons. Hadrons begin to dominate over antihadrons to produce a matter-antimatter asymmetry
Nucleo-synthesis	10^{3} sec	At the beginning of the epoch, protons and neutrons form nuclei of light elements. By the end of the epoch, larger nuclei begin to form from smaller nuclei and nucleons
Ion	300,000 yr	Temperature too hot for stable, neutral atoms to form. Universe consists of photons and ionic matter (electrons and nuclei)

The matter-antimatter imbalance is considered an asymmetry and is sketched in Figure 5.6. Modern estimates suggest that matter particles exceeded antimatter particles by approximately one matter particle per billion. The one part per billion difference between matter and antimatter is considered sufficient for explaining the preponderance of matter in the contemporary universe.

Primordial Nucleosynthesis

The scientific view of life discussed in later chapters is based on the assumption that animate creatures arise from inanimate matter. Life as we know it is based on carbon. Nucleosynthesis is the mechanism that helps us explain how atoms of living organisms were formed.

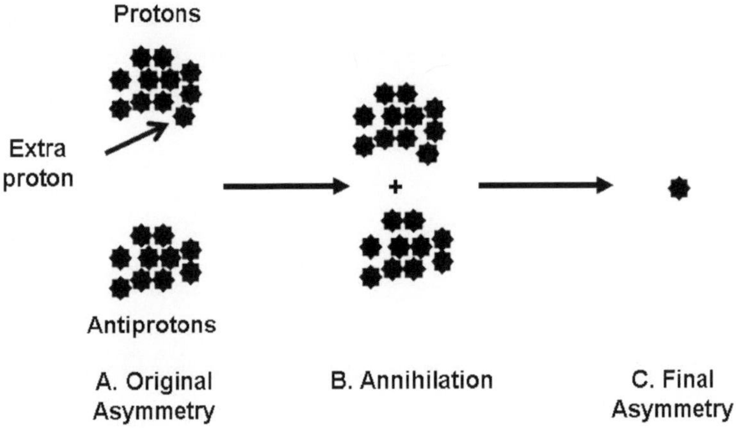

Figure 5.6. Matter–antimatter asymmetry at the Big Bang [Liddle, 1999, p. 92].

The Hadron Epoch and the matter–antimatter asymmetry were followed by nucleosynthesis. As the universe aged, it expanded and cooled. Interactions between particles began to form nuclei, atoms, and molecules. Nucleosynthesis is the formation of complex nuclei from protons and neutrons. There are two types of nucleosynthesis: primordial nucleosynthesis and stellar nucleosynthesis. Primordial nucleosynthesis is the formation of nuclei from protons and neutrons in the early universe. Stellar nucleosynthesis is the formation of nuclei in the interior of stars or by the explosion of stars. Stellar nucleosynthesis is an ongoing process that is discussed later in the next section.

Primordial nucleosynthesis occurred early in the life of the universe and was the beginning of nuclear formation. Relativistic neutrons and protons were formed by the combination of quarks during the first few minutes of the early universe. The temperature of the universe was too hot to allow the formation of stable nuclei. As the temperature cooled, one neutron and one proton could form a stable nucleus of deuterium. Two deuterium nuclei could then combine to form a tritium nucleus with a proton and two neutrons. Both deuterium and tritium are hydrogen isotopes. Once deuterium and tritium were available, they could interact to form the helium nucleus. The helium nucleus could also be formed when the light helium isotope consisting of two protons and one electron interacted with the deuterium.

The abundance of helium in the universe is attributed to primordial nucleosynthesis. Hydrogen is the most abundant element in the universe, followed by helium. All other elements are much less abundant than hydrogen and helium.

The Matter-Dominated Era

The dominant density in Figure 5.5 crosses over from radiation density to matter density in the Matter-Dominated Era. Table 5.4 presents key events in the Matter-Dominated Era. Again, recall that cosmic times in the table are approximate and subject to revision, and the increase in cosmic time is accompanied by universal expansion and cooling of the temperature of the universe.

Stellar Nucleosynthesis

The density of matter at the beginning of the Matter-Dominated Era was large enough to restrict the passage of light from one point of the universe

Table 5.4 Matter-Dominated Era (1 Ma = 1 Mega-annum = 1 million years; 1 Ga = 1 Giga-annum = 1 billion years).

Epoch	Cosmic Time (end of Epoch)	Comment(s)
	300,000 years	End of Radiation-Dominated Era
Recombination	380,000 years	Neutral atoms form. Radiation decouples from ordinary matter and the universe is transparent to photons. The CMB radiation originates during this epoch
Dark Ages	200 Ma	Ordinary matter becomes enclosed in astronomical dark matter structures. Gravity of matter amplifies primordial fluctuations into small amplitude ripples in spacetime
Galaxy Formation	1.4 Ga	Gravity continues to amplify ripples in spacetime and leads to the formation of the first galaxies
Age of Stars and Galaxies	9.8 Ga	Stars and galaxies form

to another. Since light is absorbed over relatively short distances compared to the size of the expanding universe, the universe appeared to be opaque. As the universe expanded, the density of matter decreased until it was small enough to allow the passage of light over great distances. The universe became transparent during the Recombination Epoch.

Most matter existed as stable elemental hydrogen or helium as the temperature of the universe cooled. Small perturbations, or disturbances, in the distribution of matter were growing in significance. Gravitational attractions of locally dense matter distributions were gaining strength relative to the other three fundamental interactions as distances between accumulations of matter increased. Regions of relatively strong gravitational attractions are believed to be the seeds from which galaxies sprouted some 1 billion years after the Big Bang.

Primordial nucleosynthesis was able to create the first nuclei, such as hydrogen, deuterium, tritium, helium, and, to a lesser extent, lithium. The formation of stars from primordial matter was an important step in building larger nuclei. Stellar nucleosynthesis is the process that has created and continues to create, the elements we observe in nature and classify in the Periodic Table. This material is starstuff: it is material that was formed during the lifetime of a star.

Stellar nucleosynthesis is a name that encompasses a set of processes. The first process is the creation of relatively small nuclei by nuclear fusion in the stellar interior. In this process, nuclei with the number of protons comparable to iron (56) or less are created. For example, carbon, the essential element for life as we know it, is formed in this process. To create larger nuclei, more energy and more neutrons are needed. This set of conditions occurs when some stars begin to die.

As stars age, the fusion process consumes available fuel such as helium and carbon and eventually reaches a stage when the repulsive pressure associated with nuclear fusion is overcome by gravitational forces associated with the mass of the star. Depending on the size and composition of the star, it will collapse and may experience an explosive fate by becoming a nova or supernova. The explosive gravitational collapse of a star is a stage in the formation of nuclei with approximately 60 protons or more. The details of stellar nucleosynthesis are beyond the scope of this book.

Formation of Celestial Objects

The end of the Age of Stars and Galaxies in the Matter-Dominated Era coincided with the formation of the Solar System and the origin of life on Earth approximately 4.5 billion years ago. Local accumulations of matter exert a stronger gravitational attraction than do less dense neighboring regions of space. An increase in gravitational attraction provides a mechanism for enlarging the local accumulation of matter. Figure 5.7 illustrates this effect by showing the increase in mass density from initial to final conditions around two points A and B. The gravitational capture of matter further increases the total gravitational pull of the growing accumulation. This snowballing effect results in a universe with vast distances separating regions containing substantial quantities of matter.

Regions of space having relatively dense concentrations of matter first appear as gaseous clouds adjacent to regions with less dense concentrations of matter. Gaseous nebulae, such as the Eta Carinae and Horsehead Nebulae, are modern examples of these regions. In time, local disturbances of the matter distribution within the gaseous clouds cause further coalescing of matter to form stars. All primordial matter does not necessarily coalesce into stars. Some locally dense gaseous clouds are neither massive enough nor energetic enough to initiate the nuclear forces necessary to balance the gravitational forces present in the coalescing cloud. Gaseous clouds which can establish a balance between nuclear repulsion and gravitational attraction become stars.

Hydrogen is the primary fuel for nuclear fusion reactions: two atoms of hydrogen are fused together in the dense, hot core of the coalescing cloud

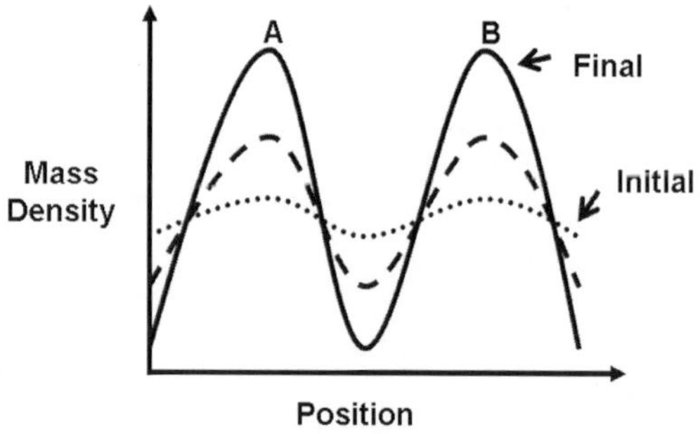

Figure 5.7. Local accumulation of matter.

to form helium. Not all of the mass-energy of the two hydrogen atoms is needed to form helium. The excess is converted to EM energy and is a major source of solar radiation. As a star ages, it burns up its atomic hydrogen fuel. In the process of burning hydrogen, the core of a star grows and its pressure and temperature increase. Eventually, the pressure and temperature of the core are high enough to begin burning helium. These nuclear reactions — the burning of hydrogen and helium — are important processes in the construction of larger atoms such as iron and magnesium.

In addition to its original composition, the destiny of a star depends on such factors as its initial size and mass. As stars near the end of their lifetime, which corresponds roughly to the time when their atomic fuel is exhausted, they may undergo several fates. Some stars explode. By-products of the explosion — which may be a nova or supernova — include many heavy elements as debris. The debris of a supernova can form a nebula from which other stars may evolve. A supernova explosion is believed to be the source of material from which our Sun and its planets, including Earth, formed. Other stars simply contract as they cool and become white-dwarfs. Still other stars undergo a gravitational collapse under their own weight until they enter a very dense state called the neutron-star state. In a few instances, black holes are the end result of the gravitational collapse of a star.

Some of the bodies that are formed by the coalescence of matter in nebular clouds are not dense enough to ignite nuclear reactions. They may be captured by stars and become satellites, or planets, of the star. Still smaller disturbances of the matter distribution in a nebula may be captured by the planets and become moons. Yet, other disturbances become asteroids, comets, and meteors.

The Kant-Laplace Hypothesis

The above discussion is a sampling of the rich variety of fates that can befall primordial matter as the universe matures. It is a scenario stemming from ideas first suggested by the German philosopher Immanuel Kant (1724–1804) in 1755 and, independently, by the French mathematician Pierre Laplace (1749–1827) in 1796. The Kant-Laplace hypothesis, also known as the nebular hypothesis, is the basis of most modern theories of nebular, galactic, stellar, and planetary formation theories.[4]

Pieces of evidence supporting the Kant-Laplace hypothesis include observations by the Infrared Astronomical Satellite (IRAS). IRAS was launched on a voyage to scan the sky in search of infrared (heat) radiation on January 25, 1983. IRAS found clouds of gas and dust containing newly formed stars and evidence of solid objects forming around stars.

The interstellar medium contains atomic nuclei, molecules, and grains of "interstellar dust." Approximately 90% of the nuclei are hydrogen nuclei (H), and 9% are helium nuclei (He). The rest of the atomic nuclei include nuclei of carbon (C), oxygen (O), neon (Ne), sodium (Na), magnesium (Mg), aluminum (Al), silicon (Si), sulfur (S), calcium (Ca), iron (Fe), nickel (Ni), etc. Molecules in the interstellar medium include hydrogen (H_2), carbon monoxide (CO), methane (CH_4), ammonia (NH_3), hydrochloric acid (HCl), silicon oxide (SiO), sodium chloride (NaCl), water (H_2O), sulfur dioxide (SO_2), hydrogen sulfide (H_2S), and potassium chloride (KCl). More complex molecules in the interstellar medium include methanol (CH_3OH) and acetone $(CH_3)_2CO$. Notice that both inorganic and organic molecules are present in the interstellar medium. The presence of organic molecules in space has encouraged some people to believe in the extraterrestrial origin of life. Interstellar dust includes grains of material similar to soot from a candle. The material in interstellar dust consists of inorganic substances such as carbon and silicates, as well as icy and organic components.

Two more pieces of evidence that favor the Kant-Laplace hypothesis of planetary formation are provided by the planets in our Solar System. The first piece of evidence comes from observations of planetary orbits. All planetary orbits lie in a plane called the ecliptic plane. An ecliptic plane is a plane that intersects the Sun and is illustrated in Figure 5.8. The ecliptic plane is not perfectly flat because the angle of inclination of the orbit of the innermost planet Mercury is 7 degrees relative to the ecliptic plane. When Pluto was considered the outermost planet, its orbit was tilted at an angle of about 17 degrees from the ecliptic plane. In 2006, the International Astronomical Union (IAU) changed the classification of Pluto from a planet to a dwarf planet.[5]

The second piece of evidence comes from observations of the direction of rotation of planetary orbits relative to the direction of rotation of the Sun.

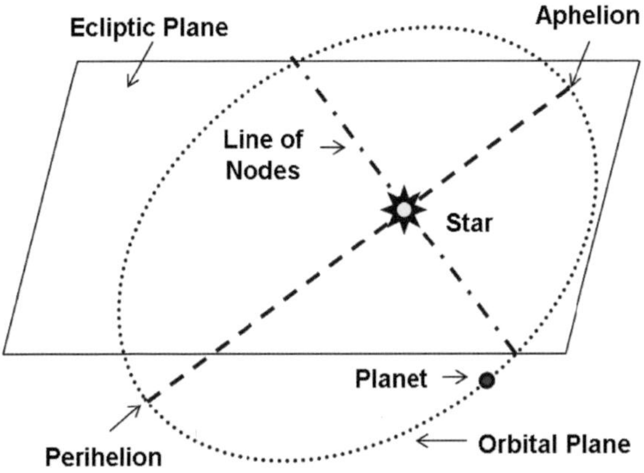

Figure 5.8. Planetary orbit and the ecliptic plane.

Our Sun rotates about its axis much like a spinning top. All planets of the Solar System orbit in the same direction as the direction of rotation of the Sun. These two observational facts — the direction of orbital rotation, and orbits confined to the ecliptic plane — suggest that the Sun and planets were formed from the dust and debris of a rotating gas cloud.

The Kant-Laplace hypothesis is considered a single-body theory because it does not require a collision between celestial objects. Alternatives to the single-body Kant-Laplace hypothesis are second-body theories. Second-body theories employ the near approach of two celestial objects as the primary mechanism for forming a third celestial object. It is possible that many mechanisms played an important role in various regions of the universe. The Kant-Laplace hypothesis continues to be a leading mechanism for explaining the formation of celestial bodies.

The Vacuum-Dominated Era

The acceleration of the expansion of the universe is associated with the cross-over from the Matter-Dominated Era to the Vacuum-Dominated Era. The cross-over occurs when the mass density of vacuum energy exceeds the mass density of matter. Table 5.5 presents key events in the Vacuum-Dominated Era. Cosmic times in the table are approximate and subject to revision.

Table 5.5 Vacuum-Dominated Era (1 Ga = 1 Giga-annum = 1 billion years).

Epoch	Cosmic Time (end of Epoch)	Comment(s)
	9.8 Ga	End of Matter-Dominated Era
Dark Energy	13.8 Ga (today)	Dark energy density exceeds the matter density and universal expansion accelerates

Point to Ponder: Is the Cosmological Timeline Correct?

A review of published timelines[6] shows a general sequence that proceeds from radiation-dominated to matter-dominated to vacuum dominated, but details differ substantially within those eras. It is very unlikely that the timeline in Tables 5.3 through 5.5 is correct. I have presented it as an example of a cosmological timeline that is still being studied.

❁ ❁ ❁

We have seen in this chapter that the quantum foam is considered a plausible scientific explanation of the initiation of the Big Bang. A cosmological history of the physical universe based on inflationary cosmology has been presented. The fate of the universe depends on whether the universe is flat, open, or closed. If the universe continues to expand, stars will eventually burn out and fewer stars will be born to replace them in a darkening universe. It is now time to consider the role of life in this scenario.

Chapter 6

What Is Life?

The observations and associated theories described in preceding chapters have set the stage for a scientific description of the origin and evolution of life. To understand the origin of life,[1] we must understand the chemical basis of life.[2] We then consider the question: what is life?

Models of the Atom

Chemists in the 18th and 19th centuries made a distinction between chemicals of inanimate material and chemicals of living organisms. They called the former inorganic, and the latter organic. A distinction was considered necessary because post-Renaissance chemists were unable to make organic chemicals from inorganic materials. Although the distinction is still made today, it is used as a matter of tradition rather than necessity. The distinction was rendered unnecessary in 1832 by German chemist Friedrich Wöhler (1800–1882). Wöhler synthesized the organic compound urea by heating the inorganic compound ammonium cyanate. Wöhler's work is a milestone of chemistry and a precursor of molecular biology.

Chemistry is the study of relationships between atoms and molecules. Molecules are groupings of atoms, and atoms consist of negatively charged electrons bound to a positively charged nucleus by the electromagnetic interaction. The nucleus contains positively charged protons and electrically neutral neutrons bound together by the strong interaction. The nucleus occupies a small volume of space relative to the volume occupied by the atom.

The first modern model of the atom was developed by Niels Bohr (1885–1962) in 1913. Bohr envisioned electrons orbiting nuclei in direct analogy with planets orbiting the Sun. Bohr's atomic model was based on an *ad hoc* synthesis of classical and early quantum concepts. His model agreed with many experimental results, particularly those of atomic spectroscopy, but it contained assumptions which violated then accepted theoretical principles. For example, classical physics said that an orbiting electron should emit electromagnetic radiation, which would make atoms unstable. Bohr's work on atomic structure came at a time when the flaws of classical physics could no longer be ignored.

Modern models of atomic structure are based on quantum principles and have many similarities to Bohr's model. Electrons still occupy most of the volume of the atom, and nuclei are still surrounded by electrons. The positions of electrons are not known with certainty, however, in accord with Heisenberg's uncertainty principle. Bohr's electron orbits are now viewed as probabilistic electron orbitals. The orbitals allow us to determine the probability of locating an electron at various positions around the nucleus. Inner electrons occupy orbitals close to the nucleus, while outer electrons occupy orbitals farther from the nucleus. The collection of orbitals can be viewed as an electron cloud and represents the probability of finding electrons at different locations within the atom. Typical shapes of electron clouds are depicted as dashed lines in Figure 6.1.

Electron energies are quantized in both Bohr's model and modern atomic models. In retrospect, the success of Bohr's atomic model was due in large part to the incorporation of quantized electron energy levels in his model. Some energy levels are sketched as horizontal lines in Figure 6.2. These levels represent the only energies an electron can have in an atom. No other energy values are allowed by quantum theory. Experimental measurements support this theoretical requirement.

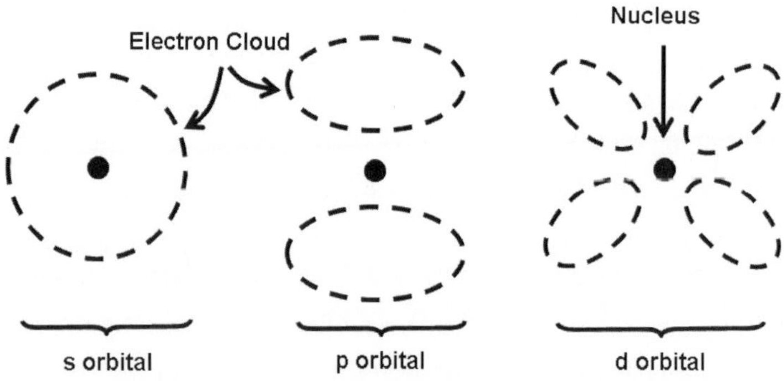

Figure 6.1. Typical shapes of electron clouds.

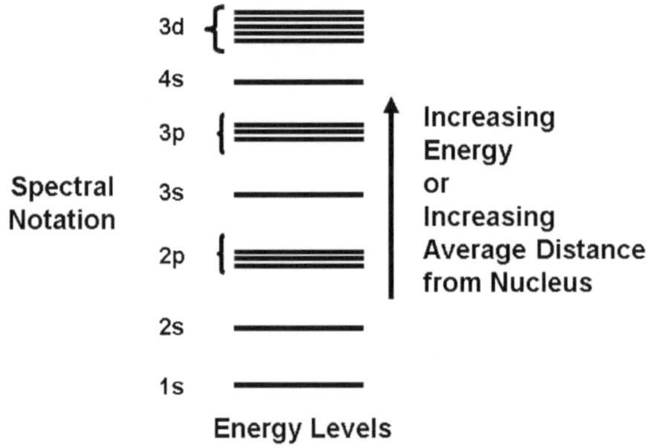

Figure 6.2. Atomic energy levels.

The identity of an element is based on the number of protons in the nucleus. An electrically neutral atom must have as many negatively charged electrons in its electron cloud as it does positively charged protons in its nucleus. Atomic electrons fill energy levels with the lowest energy first. Only two electrons are allowed in each energy level. The restriction to two electrons is quantum mechanical in origin.

Electrons behave in the presence of magnetic fields as if they were spinning like tops. This is a simplistic concept, however, because electrons are treated mathematically as point particles with a negligible physical volume. The analogy of a spinning top to electron spin helps convey the idea of electron spin, which was first suggested by German physicist Otto Stern (1888–1969) in 1921. A year later Stern and his countryman, Walther Gerlach (1889–1979), experimentally demonstrated the existence of electron spin. They observed a beam of silver atoms split into two beams by an inhomogeneous magnetic field.

Dutch-American physicists Samuel A. Goudsmit (1902–1978) and George E. Uhlenbeck (1900–1988) explained the Stern-Gerlach results by applying the concept of electron spin. Only two values of electron spin are possible. The two values are referred to by many names: spin up and spin down; clockwise and counterclockwise spin; and helicity. Helicity is the most general concept. Helicity defines the direction of a particle's spin relative to the direction of motion of the particle. Electron spin is one of four quantum numbers characterizing the behavior of an electron in an atom. The other three quantum numbers define the spatial configuration of the electron cloud.

The most important quantum number, the principal quantum number, represents the average distance of an electron from the atomic nucleus. The shape of the electron orbital is characterized by the secondary quantum number. Three examples of secondary quantum numbers are shown in Figure 6.1. Secondary quantum numbers are represented by letters. This practice is a relic of early spectroscopic notation. The first few letters are s, p, d, and f. They signify secondary quantum numbers with absolute values of 0, 1, 2, and 3 respectively. Figure 6.2 shows the principal and secondary quantum numbers of the lowermost electron energy levels. For example, the spectral notation 2p denotes a principal quantum number of 2 and a secondary quantum number with an absolute value of 1.

As the principal quantum number increases, the average distance of the electron from the nucleus increases. The secondary quantum number defines the shape of the electron orbital. A third quantum number, called the magnetic quantum number, defines the orientation of the electron cloud in space. The s orbital, for example, is a spherical distribution. By contrast, the p orbital is split into three energy levels because it can be aligned along the x, y, or z axes. Five unique orientations are possible for the d orbital, and are depicted by the five energy levels shown in Figure 6.2. Figure 6.3 adds the magnetic quantum number to the spectral notation of the 2p orbital.

	1s	2s	$2p_x$	$2p_y$	$2p_z$
Hydrogen	↑				
Helium	↑↓				
Lithium	↑↓	↑			
Berylium	↑↓	↑↓			
Boron	↑↓	↑↓	↑		
Carbon	↑↓	↑↓	↑	↑	
Nitrogen	↑↓	↑↓	↑	↑	↑
Oxygen	↑↓	↑↓	↑↓	↑	↑
Fluorine	↑↓	↑↓	↑↓	↑↓	↑
Neon	↑↓	↑↓	↑↓	↑↓	↑↓

↑ Electron spin up
↓ Electron spin down

Figure 6.3. Simple atoms.

Experimentalists have found that every electron in an atom must have a unique set of quantum numbers. This observation is profoundly related to the statistical behavior of many identical particles. A discussion of this point would lead us too far afield for our purposes. Suffice it to say that no two electrons can have the same set of quantum numbers in an atom. Every electron orbital contains no more than two electrons, and the two electrons in each orbital have different values of the spin quantum number. Electron orbitals for the first few elements are shown in Figure 6.3. The electron configuration of helium, for example, is the same as hydrogen plus one electron added to complete the 1s orbital. Electron configurations of each succeeding atom are constructed by adding an electron to the lowest energy, unoccupied electron orbital. The electron configuration of every known atom can be understood using this prescription.

Point to Ponder: What Happened to the Quark Model?

The discussion of atomic models presented above is largely told from the point of view of particles rather than fields. From a more modern perspective, electron fields are interacting with electrically charged quarks confined in nucleons (protons and neutrons). Each nucleon contains three quarks. The strong interaction between quark fields binds the nucleons. It is mathematically and conceptually simpler to describe atomic models using the more traditional particle perspective.

Molecular Biology

The energy of an atom is a minimum when all of its outermost electron orbitals are filled. Helium and neon in Figure 6.3 are examples of atoms with filled electron orbitals. Both of these elements are chemically inert. To achieve this state, atoms with partially filled outer electron orbitals can share electrons, give up electrons, or take electrons from other atoms. For example, hydrogen can give up its single electron to fluorine in an explosive synthesis of the hydrogen fluoride molecule. Two hydrogen atoms and one oxygen atom can combine to form a water molecule. Table salt, sodium chloride, is a molecule with one atom of sodium and one atom of chlorine. Water and table salt represent two important types of bonds between atoms in a molecule: ionic bonds and covalent bonds. These bonds are illustrated in Figure 6.4.

Sodium chloride bonds are ionic: sodium gives up one electron and becomes positively charged while chlorine accepts the sodium electron and becomes negatively charged. The opposite charges of the ionized sodium and chlorine atoms result in an electromagnetic attraction. This attraction is the ionic bond.

The bonds between oxygen and hydrogen atoms in water are covalent: the atoms share electrons when their orbitals overlap. This overlap represents the probability of finding a shared electron in the orbital of its host atom or in the orbital of the bound atom. Organic molecules, the molecules of life, rely on covalent bonds.

Examples of simple molecules are shown in Figure 6.5. A molecular bond is denoted by a dash between two bound atoms. Each dash represents the sharing of two electrons. Sometimes the molecular bond is formed by the sharing of two or three electron pairs. These multiple bonds are referred to as double and triple bonds, and are depicted by two and three dashes, respectively. The oxygen molecule and the nitrogen molecule are examples of molecules with double and triple bonds.

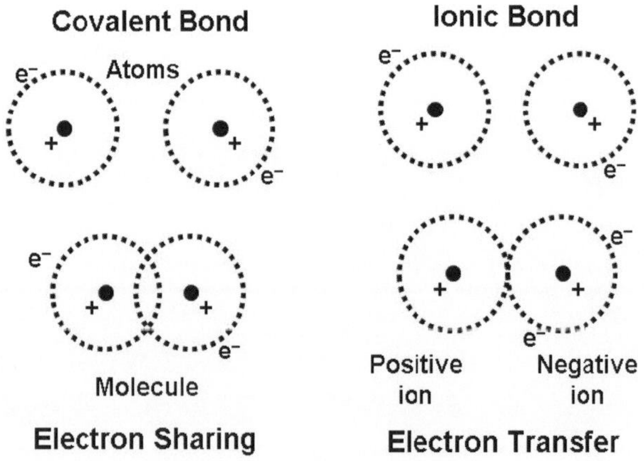

Figure 6.4. Molecular bonds.

In addition to introducing molecular symbolism, Figure 6.5 includes sketches of two molecules containing carbon. Examples of more complex carbon-bearing molecules are shown in Figure 6.6. The letter "R" in the glycine base is a placeholder for an amino acid shown in Figure 6.7 and discussed as follows. The basic difference between inorganic and organic chemicals is that organic chemicals are compounds of carbon C. Biochemistry, the chemistry of life, is a branch of organic chemistry, the chemistry of carbon compounds. Molecular biology is a study of the chemistry of life on the molecular level.

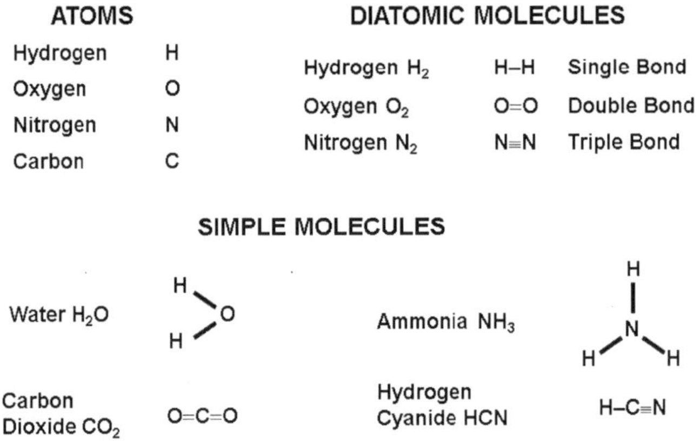

Figure 6.5. Simple molecules.

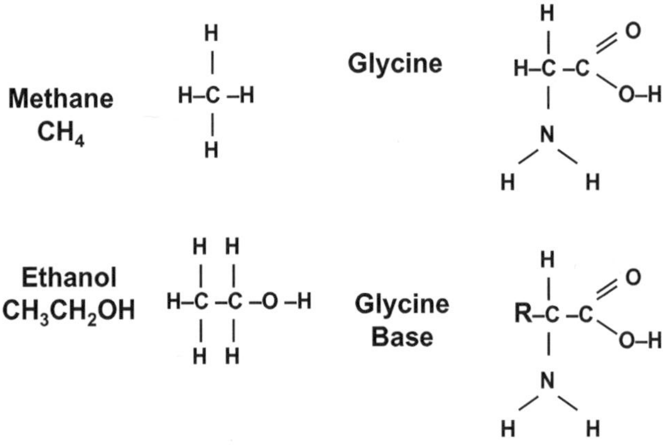

Figure 6.6. Simple organic molecules.

Cells

The word "cell" originated with English physicist Robert Hooke (1635–1703) in 1665. Using a microscope, he observed regularly shaped structures in thin sections of cork. These structures he called cells. Today we know that Hooke's "cells" were only cellulose walls. Modern usage of the word "cell" refers to the basic unit of an organism. The cell is a biochemical factory that is encased in a porous membrane. This connotation is a consequence of the work of German botanist Matthias J. Schleiden (1804–1881) and countryman Theodor Schwann (1810–1882), a zoologist.

Schleiden and Schwann postulated that all living systems, from microscopic organisms to humans, are composed of cells. Their postulate has been borne out by observation. Every living system is constructed from cells; the chemical contents of every cell are contained within a boundary called the cell wall or cell membrane; and every cell contains chemicals of great size and complexity known as proteins and nucleic acids.

Two of the most important classes of biochemicals are proteins and nucleic acids. Proteins are built up from a set of twenty smaller molecules called amino acids. Figure 6.7 illustrates some typical amino acids. Except for

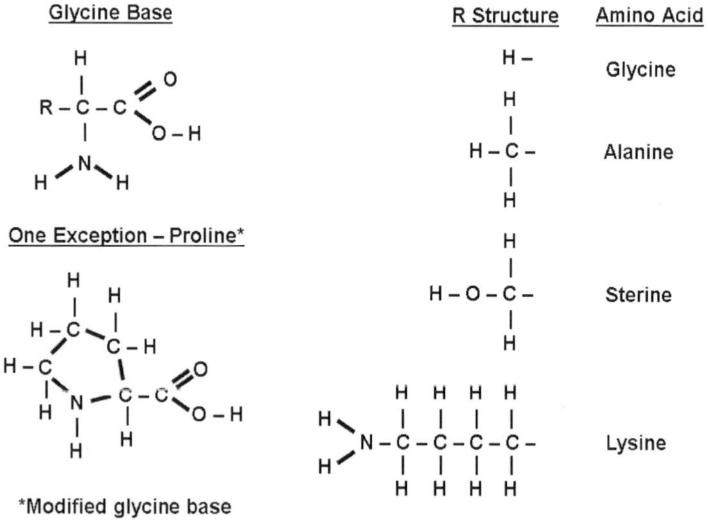

Figure 6.7. Typical amino acids.

proline, every naturally occurring amino acid has a glycine base. Substitution of a glycine hydrogen with other organic compounds is responsible for the various amino acids. A bond between two amino acids can be obtained by combining a hydroxyl (OH) group from one amino acid with a hydrogen atom from another amino acid. Repetition of this process leads to the formation of long chains of amino acids. These chains, or polymers, of amino acids are proteins. Enzymes are an important class of proteins. They function as catalysts in the cell. As such, enzymes can speed up a chemical process without being changed by the process. Enzymes are necessary for the replication of nucleic acids.

The nucleic acids deoxyribonucleic acid (DNA) and ribonucleic acid (RNA) are essential for all known forms of life. DNA is a large molecule formed from combinations of four smaller molecules attached to a sugar (deoxyribose phosphate) base. A sketch of a DNA molecule is presented in Figure 6.8. DNA exists as a double helix. Each helical strand is a chain of deoxyribose phosphate molecules. Attached to the helical strand are organic molecules known as purines (adenine (A) and guanine (G)) and pyrimidines (thymine (T) and cytosine (C)). A purine can pair with a pyrimidine to form a base pair that bonds two helical strands of a DNA double helix.

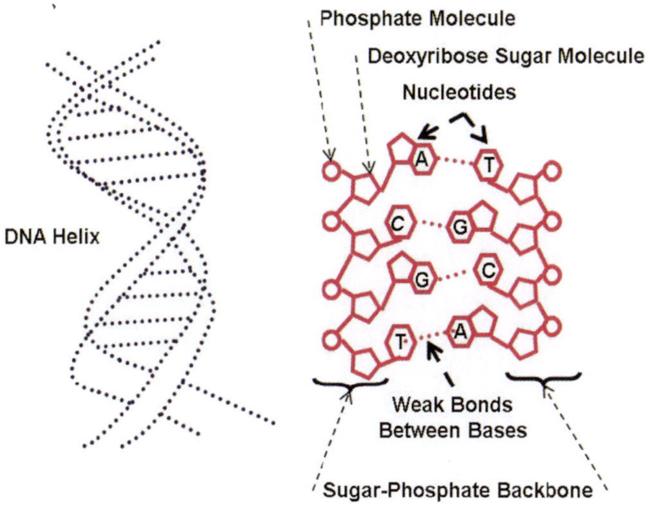

Figure 6.8. DNA.

An encoded sequence of DNA base pairs that forms the rungs of the double helix is a genome. Modern genomes appear to have evolved from ancestral cells by the gradual mutation of DNA sequences. Differences between DNA sequences, or genomes, are associated with evolutionary branching. Molecular phylogeny is a comparative study of genomes and the relationship between genome differences and evolutionary traits.

Nucleic acid is the genetic material in the cell.[3] The nucleic acid codes the function of the cell and controls reproduction. Ribosomes in the cell translate the genetic code carried by nucleic acids into proteins. The rules governing nucleic acid bonding are known as the Watson-Crick rules after their discoverers, English biophysicist Francis Crick (1916–2004) and American molecular biologist James Watson (1928–). Crick and Watson were the first to deduce the molecular structure of the DNA molecule. They found that adenine only attached to thymine, and guanine only paired with cytosine. The spatial ordering of adenine-thymine and guanine-cytosine pairs along the DNA structure is the identifying characteristic of a DNA molecule. The RNA molecule is the template by which DNA replication occurs.

RNA is a single helical structure. Unlike DNA, which is found primarily in the nucleus of a cell (Figure 6.9), RNA exists principally in the protein synthesizing ribosomes. The chemical content of RNA is similar to DNA. In addition to assisting the replication of DNA during cell division, RNA also plays an important role in the synthesis of cellular proteins. Cellular

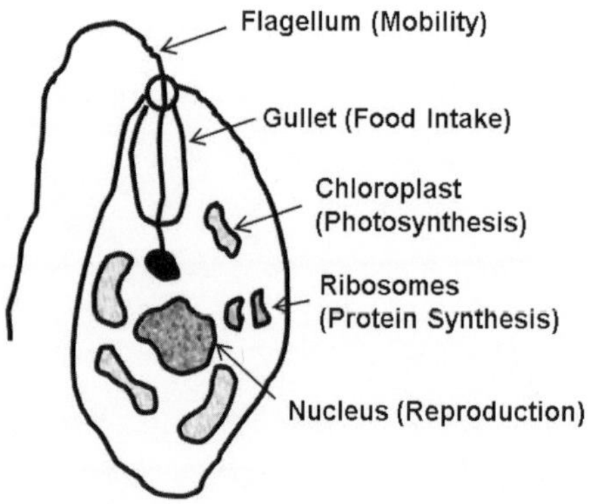

Figure 6.9. Unicellular organism.

proteins are synthesized according to instructions in the DNA molecule. The DNA information cannot be used directly in protein synthesis; it must be communicated by an intermediary. RNA is the intermediary. The relationships between DNA, RNA, and proteins form the central dogma of molecular biology. As summarized by Rob Phillips and Stephen Quake, the central dogma "refers to the set of processes whereby DNA is copied (replication), genes are read and turned into messenger RNA (transcription), and finally, messenger RNA is turned into the corresponding protein by ribosomes (translation)" [Phillips and Quake, 2006, p. 41].

The simplest form of life appears to be bacteria because bacteria contain both DNA and RNA. As we have seen, DNA replication requires RNA as an intermediary. The presence of both DNA and RNA in bacteria means that bacteria can reproduce without any external assistance. Viruses cannot reproduce without the help of a host organism. It is not clear whether viruses played a role in the development of the first unicellular organisms.

Reproduction of a unicellular organism, such as the ancestral flagellate sketched in Figure 6.9, requires accurate replication of cellular DNA. Mutations, or mistakes in the replication of DNA, can occur. These mistakes are necessary for changes in a unicellular species to occur. More complex, sexually active organisms are constantly making new combinations of genetic material. These new combinations, and mistakes in the accurate replication of combinations, provide mechanisms for the evolution of sexual organisms. The same mechanisms do not exist for asexual organisms. Asexual unicellular organisms need mutations to change and evolve. Thus, variations in DNA replication are essential to a viable theory of evolution from simple unicellular organisms to complex multicellular organisms.

Definitions of Life

The need for a definition of life has acquired new urgency as humanity has learned more about the reproductive process and as we extend our reach to other planets. How will we know if life exists elsewhere in the universe if we do not know how to define life? How can we have confidence in laws that govern the reproductive rights of human beings, especially women and the unborn, when we do not know what life is?

Life is difficult to define.[4] British chemist Leslie Orgel (1927–2007) defined a living entity as a CITROENS, a Complex, Information-Transforming Reproducing Object that Evolves by Natural Selection. A structure is considered living if it satisfies the following criteria: the object is complex and yet well-specified, such as DNA; and the object can reproduce or is descended from reproducing objects.

In his search for life in the Solar System as part of the National Aeronautics and Space Administration, American geneticist Norman Horowitz (1915–2005) presented a definition of life based on genetics. Horowitz[4] said that "Life is synonymous with the possession of genetic properties." He recognized two important genetic properties: self-replication, and mutation. In response to the questions of who or what is the designer of the living world, Horowitz said the designer is "the cumulative effects of natural selection acting on spontaneous mutations over long periods of time."

Noam Lahav reviewed several historical definitions of life and adopted a "scientific" view. Lahav[4] said that "Living entities are complex, far-from-equilibrium structures maintained by the flow of energy from sources to sinks." According to Lahav, living entities can replicate, mutate, exchange matter and energy with their environment, and evolve.

Richard Cohen identified several conditions that he considered necessary for the origin of life.[4] Energy is needed to form complex organic molecules. Once formed, complex organic molecules need protection from strong radiation to prevent the disassociation of the molecule. The protection can be provided by several media, such as water, ice, or rock. The chemicals needed for life need to be concentrated. This can occur, for example, in a pool of water that evaporates after the chemicals of life have formed. Finally, catalysts are needed to facilitate the chemical reactions needed for the origin of life. Enzymes are organic catalysts.

Horst Rauchfuss[4] provided a summary of definitions of life formulated by the NASA Exobiology Program. The goal of the NASA Exobiology Program is "to understand the origin, evolution, distribution, and future of life in the Universe" [NASA Exobiology, 2020]. It operates five areas of research:

- Planetary conditions for life: seek "to delineate the galactic and planetary conditions conducive to the origin of life."
- Prebiotic evolution: seek "to understand the pathways and processes leading from the origin of planetary bodies to the origin of life."
- Early Evolution of Life and the Biosphere: "determine the nature of the most primitive organisms and the environment in which they evolved."
- Evolution of Advanced Life: "determine the biological and environmental factors leading to the development of multicellularity on Earth and the potential distribution of complex life in the Universe."
- Exobiology for Solar System Exploration: relate "relating what is known about life on Earth to conditions prevailing on other planetary bodies."

These programs show that there is still much to learn.

Criteria for Life

As we have seen, the criteria for life are not well-defined. A set of criteria can be selected from the set of characteristics common to living systems.

- To be considered living, an entity must be able to reproduce. If the entity is sexually sterile, as mules are, the entity must be descended from other entities which are able to reproduce.
- Mistakes or genetic variations associated with the reproductive process are the mechanisms by which mutations occur. Mutations, in turn, are necessary for evolution.
- Another behavioral characteristic of a living system is the ability to acquire and process food. Food in this sense can be as simple as molecules or as complex as other organisms.
- A necessary part of the metabolic process, in addition to the acquisition and processing of food, is the removal of waste products. Every biological system can excrete waste products.
- A final trait is the ability of biological systems to respond to external stimuli. The degree of sophistication of actual responses depends on the type of external stimulus, and the complexity of the responding organism.

If we adopt these criteria as a reasonable set of criteria for recognizing life, we can consider the question: how did life originate?

❁ ❁ ❁

The origin of the simplest forms of life depends on several factors. Macromolecules need to be formed from smaller molecules by carbon fixing, or the addition of carbon atoms. Once formed, macromolecules must be able to perform several functions, such as serve as a cell-like enclosure with an inside and an outside, convey information, and be self-replicating. Table 6.1 summarizes a plausible evolutionary sequence for the development of life from inorganic chemicals. It depicts an increasing level of chemical complexity as we move from inorganic chemicals to the simplest unicellular organisms, but many questions remain. For example, scientists do not know whether the chemistry of the origin of life required improbable reactions working over vast intervals of time, or probable reactions working over a relatively short period of time.

Scientists generally believe that the sequence depicted in Table 6.1 is essentially correct. This view of the transition to life does not require a supernatural entity, but it does require faith that science will eventually fill in the gaps in our knowledge as we move from one step in Table 6.1 to another. Our presentations of the origin of life in Chapter 7 and the evolution of life in Chapter 8 are designed to display the state of the art of the scientific view.

Table 6.1 Summary of transition to life.

Step	Entity
1	Inorganic chemicals
2	Organic chemicals
3	Nucleic acids and proteins
4	Prokaryotes (no nucleus; e.g. bacteria)
5	Eukaryotes (nucleic acid in nucleus; e.g. amoeba)

Chapter 7

The Origin of Life

Our review in the previous chapter has revealed two key observations: carbon-based life exists on Earth, and all known living organisms are based on cell structure.[1] Organisms range in complexity from single all-purpose cells to collections of millions of specialized cells. The simplest living organisms consist of a single cell. These unicellular organisms are believed to be the ancestors of all life. For this reason, British chemist Leslie E. Orgel (1927–2007) said that "asking how life began is equivalent to asking how the first unicellular organisms evolved" [Orgel, 1973, p. 34]. How did the first unicellular organism originate? We consider a few possible scientific explanations in this chapter, beginning with the concept of spontaneous generation.

Spontaneous Generation

Throughout most of history, people believed that inanimate systems were somehow different from living organisms. They thought that living organisms contained a quality that set them apart from inanimate objects and even the remains of once living organisms. Is life endowed with a special quality beyond our grasp, or can we create living systems from inanimate materials?

The concept of spontaneous generation — the creation of life from inanimate materials — has existed since before the age of Aristotle (ca. 384–322 BCE). Aristotle was an advocate of spontaneous generation. He believed, for example, that frogs formed from mud. His views were widely accepted for two thousand years. The first serious attack on the concept of spontaneous generation was mounted by Italian physician Francesco Redi (1626–1697) in 1668.

Prior to Redi's work, people thought that maggots formed from decaying red meat. Redi laid a piece of fine meshed gauze on the exposed surface of a piece of red meat. The gauze prevented the eggs of insects from settling on the meat. As long as the eggs of insects were isolated from the meat, no maggots formed. Redi's experiment challenged the validity of the concept of spontaneous generation. His work may have been more influential had a new discovery not been made by Dutch naturalist Antonie van Leeuwenhoek (1632–1723). Using a microscope of his own design, Leeuwenhoek made careful drawings and descriptions of microbes. He reported his work to the British Royal Society in a series of letters dated as early as 1674. The observation of microbes led some to suggest that these tiny creatures, which appeared suddenly in decaying food, were generated spontaneously. Although spontaneous generation was in retreat, it had not yet been defeated.

The final blow to the Aristotelian version of spontaneous generation was delivered by French biochemist and microbiologist Louis Pasteur (1822–1895). Through a series of clever experiments, Pasteur proved that microbes do not appear in decaying food unless the food is accessible to them or their parents [Rauchfuss, 2008, Section 1.2]. Pasteur reported his results in April 1864. The spontaneous generation of life, as understood by the ancient Greeks, ceased to be a viable theory.

Russian biochemist Alexander I. Oparin (1894–1980) revived the concept of spontaneous generation in 1923, but in a different form. He conceded that spontaneous generation does not occur on the modern Earth. Oparin hypothesized, however, that conditions on the primordial Earth were different from what they are today. He suggested that life could have emerged from inanimate material under the prevailing conditions of the primordial Earth. A similar hypothesis was independently proposed by British geneticist John B. S. Haldane (1892–1964). Is it possible that life arose from inanimate material on the primordial Earth? What were the conditions on the primordial Earth?

The Primordial Earth

Our home is Earth.[2] Technological advances have helped us improve our understanding of the planet we live on. Satellites and sensitive measurements of slight gravitational variations on Earth's surface let us effectively see through the oceans and construct a view of the hard outer crust of the Earth. Seismic detectors let us measure the speed of propagation of wave-like vibrations caused by earthquakes and nuclear detonations. These seismic measurements give us an indirect picture of the Earth's interior. In addition to an arsenal of powerful devices for extending our senses, we can also look into Earth's past by looking at the Grand Canyon in Arizona, examining fossil remains from Siberia, or studying drill cuttings from oil and gas wells drilled in the crust of the Earth. A wealth of observations obtained from a combination of methods practiced by an army of geoscientists over many tens of years is the raw material underlying the modern scientific understanding of the Earth and some leading models of the origin of life. This understanding is based on an application of the Kant-Laplace hypothesis to the formation of our Solar System.

The Solar System and Planetary Formation

Imagine a cloud of rotating gas and dust in space. Gravitational attraction between the constituents of the cloud is greatest near the center. As the cloud collapses, increases in pressure and temperature at the core of the cloud are eventually enough to ignite nuclear fusion processes and the Sun is born. The shape of the Sun continues to define itself as perturbations in the rotating cloud surrounding the infant star coalesce into protoplanets that eventually become the large, gaseous planets Jupiter and Saturn. Dust grains, some made of such relatively heavy elements as iron and magnesium, are aggregating in other parts of the cloud engulfing the Sun. These heavy aggregates are called planetesimals and are the precursors of the planets Mercury, Venus, Earth, and Mars. Planetesimal growth is greatest near the Sun where more massive nebular material is most abundant, yet not so close that it is pulled into the Sun.

The dense core of each planetesimal acquires an exterior of less dense material from the dwindling molecular gas cloud. Gravitational pressure from the planetesimal exterior generates a large interior pressure. The temperature also rises, in part due to crushing gravitational pressure, and in part due to nuclear reactions. Eventually, the pressure and temperature of the planetesimal interior become great enough to melt a portion of the once solid core. Moving outward from the interior we find the pressure and temperature decreasing as the amount of confining material decreases and we approach the cold of space. Heavier radioactive elements migrate toward the core, while lighter elements migrate toward the planetesimal surface. The planetesimals are large enough to retain a gaseous atmosphere of light elements including hydrogen and helium. Atmospheres of the planetesimals vary from a trace of helium in the case of Mercury to an atmosphere of molecules such as ammonia and methane. Lighter molecules, especially hydrogen, escape the gravitational pull of the planetesimals.

Embodied in this model of planetary formation based on the nebular hypothesis is the explanation of some observational facts. The cross-section of the Earth's interior is subdivided into an inner core, the central mantle, and the outer crust (Figure 7.1). Seismic measurements of earthquake waves have shown the core to consist of molten metal enveloping a solid core. The Earth's electric and magnetic properties and the density of the core provide evidence for identifying the metal as an alloy of iron and nickel. Iron is the dominant constituent. Existence of a solid inner core implies the planetesimal seed did not entirely melt. The core has a radius of approximately 2,200 miles (~3,500 km) and consists of a crystalline inner core and a molten outer core.

Exterior to the core of the Earth is the 1800-mile thick (~2,900 km) mantle. The mantle is thought to be primarily basalt, a dark volcanic rock. Basalt is composed of magnesium and iron silicates. Basalt at the surface of

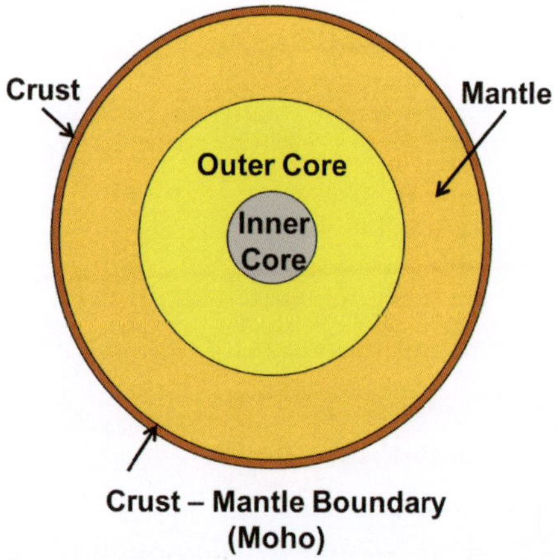

Figure 7.1. The interior of the Earth (not to scale).

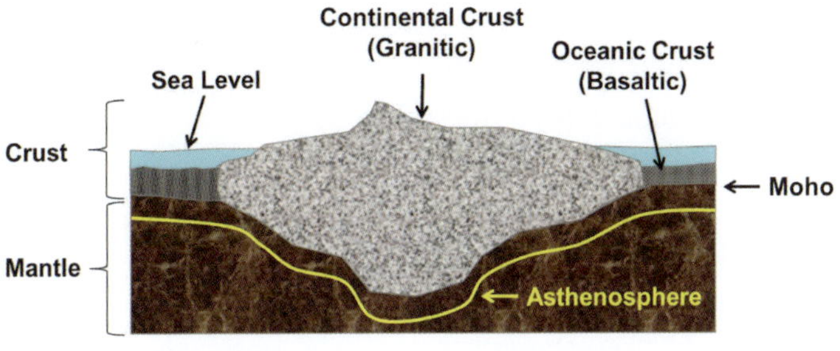

Figure 7.2. The crust of the Earth.

the mantle exists in a semi-molten state. This layer of semi-molten basalt is called the asthenosphere.

The ratio of iron to magnesium and silicon in the Earth is about the same as the ratio observed in stars like our Sun. This observation lends support to models of planetary formation that are analogous to models of stellar formation. The materials used to form the Sun and the planets appear to be from the same source.

Above the mantle is 3–50 miles (~5–80 km) of crust. Figure 7.2 is a sketch of the crust of the Earth. The granitic continental crust is composed principally of silica and aluminum; the oceanic crust contains silica and magnesia. Underlying the continental and oceanic crusts is a layer of solid basalt. The boundary between the crustal basalt and the semi-molten basalt of the mantle is the Mohorovičić discontinuity, or Moho for short. The combination of crust and solid basalt above the Moho is known as the lithosphere. We live on the lithosphere. Its history is intertwined with the history and distribution of life.

Radiometric Dating

The chronology of the lithosphere relies on a technique known as radiometric dating.[3] An element is classified by the number of protons in its nucleus. Isotopes of the element are made by adding or subtracting neutrons from the nucleus.

Most elements have stable isotopes. For example, the carbon atom has six protons in its nucleus. Carbon-12 is the isotope of carbon with six neutrons in its nucleus. The number 12 following carbon is the sum of six protons and six neutrons. Carbon-12 does not decay radioactively and is stable. An important isotope of carbon, carbon-14, has six protons and eight neutrons in its nucleus. It is an unstable isotope of carbon.

Lifetimes of radioactive isotopes are expressed in terms of half-lives. Half-life is defined as the length of time it takes half of the original quantity of an isotope to decay. The age of a fossil or rock can be estimated by combining the half-life of an isotope with knowledge of its abundance and the abundance of its decay products in a source rock. Determining the age of a rock or fossil using radioactive decay techniques is called radiometric dating.

The common occurrence of carbon-14 in living things, combined with a half-life of about 5,600 years, makes carbon-14 decay a valuable process for dating fossils. Another important decay process is uranium decay. Uranium isotopes decay by emitting helium nuclei (alpha emission). Stable decay products of uranium are lead and helium. The appearance of uranium in many rocks and their half-lives make uranium decay an important process for estimating the age of rocks. For example, the isotope Uranium-238 has

Table 7.1 Abridged geologic time scale [USGS Time, 2007].

Eon	Era	Period	Approximate Start of Interval (MYBP)
Phanerozoic	Cenozoic	Quaternary	1.8
		Tertiary	65
	Mesozoic	Cretaceous	145
		Jurassic	200
		Triassic	251
	Paleozoic	Permian	299
		Carboniferous	359
		Devonian	416
		Silurian	444
		Ordovician	588
		Cambrian	542
Proterozoic			2500
Archean			4000
Hadean			

92 protons, 146 neutrons, and a half-life of approximately 4.5 billion years, which is approximately the age of the Earth. The isotope Uranium-235 has 92 protons, 143 neutrons, and a half-life of approximately 700 million years. Rock dating is crucial to determining the age of the Earth and developing the theory of plate tectonics.

Table 7.1 is an abridged version of the geologic time scale beginning with the formation of the Earth. The most encompassing period of time is the Eon, which is subdivided into Eras, and further subdivided into Periods. The acronym MYBP in the table stands for millions of years before the present. The starting time of each interval is considered approximate since there is still uncertainty in the actual chronology of the Earth.

Plate Tectonics and Pangaea

The dynamic relationship that exists between the lithosphere and the asthenosphere has had a significant impact on the history of the lithosphere. The lithosphere consists of a set of rigid plates floating on the semi-molten asthenosphere. Molten material in the asthenosphere can enter the lithosphere through cracks between plates. This transfer of material can be by violent volcanic eruptions, or it can be gradual through the extrusion of basaltic lavas at the boundaries between plates. Volcano examples are well known. Extrusion examples are not as well-known because major extrusions were difficult to access.

Satellite measurements of the gravitational field of the Earth provide pictures of Earth's surface devoid of both oceans and vegetation. Satellite pictures clearly show boundaries between continents, and tremendous mountain ranges rising from ocean floors. The shapes of the boundaries are suggestive of vast plates, as depicted in Figure 7.3. Only the largest of the known plates are depicted in the figure. Many of these plates are associated with continental landmasses.

Marine scientists have discovered that oceanic mountain ranges are sources of basaltic extrusion and sea-floor spreading. As sea-floors spread, the continental plates are forced to move. Movement of continental plates is known as continental drift and was first proposed[4] by German meteorologist Alfred Wegener (1880–1930) in 1915. The mid-Atlantic Ridge is an example

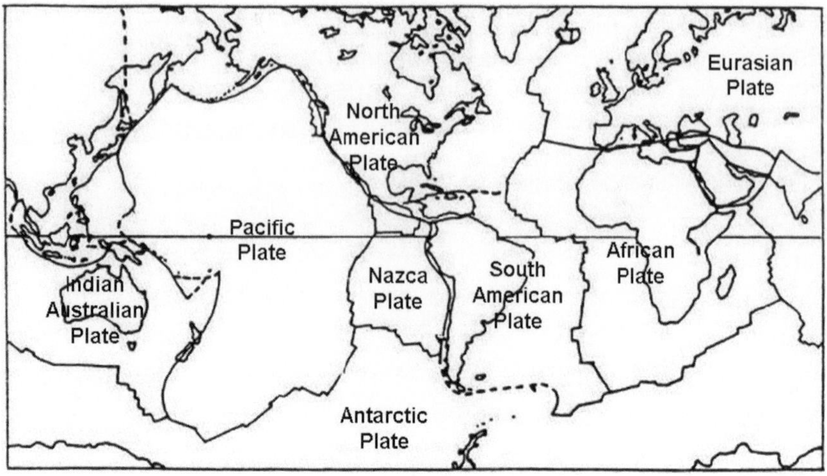

Figure 7.3. Tectonic plates.

of a subsea extrusion zone. It is a mountain range that separates the Eastern and Western hemispheres. The volcanic island Iceland is located on the mid-Atlantic Ridge. The subsea mid-Atlantic ridge includes high pressure, high-temperature zones that are unsuitable for most, but not all, carbon-based organisms. These extreme conditions are now considered a possible set of conditions for the extremophilic origin of life, as we discuss in more detail below.

Regions of some moving continental plates may collide. A head-on collision of two plates can form great mountain ranges such as the Himalayas. Alternatively, a collision can deflect one plate beneath another. Material in the deflected region, or subduction zone, may be forced down through the Moho and into the semi-molten asthenosphere. Thus, subduction zones are locations where crustal material is returned to the mantle. Together, sea floor spreading and subduction zones are the primary mechanisms for transferring material between the crust and the upper mantle. The theory describing the movement of lithospheric plates is known as plate tectonics. Using radiometric dating and similarities in geologic structures, a reconstruction of lithologic history is sketched in Figure 7.4. The terms Permian, Triassic, Jurassic, and Cretaceous are geological periods that help geologists cross-reference rocks and geologic events from one location to another.

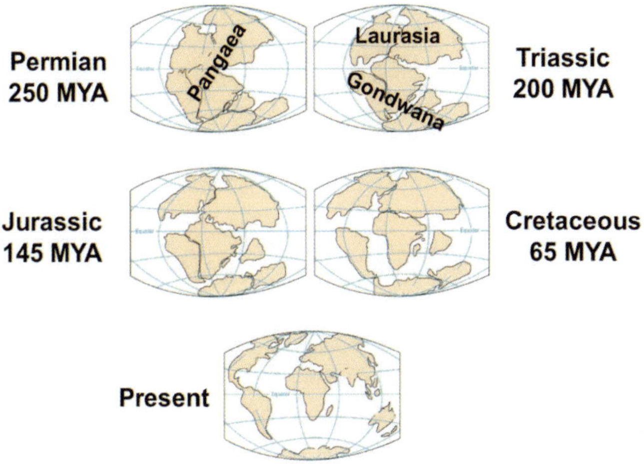

Figure 7.4. Tectonic plate movement [USGS Continental Drift, 2020] (1 MYA = 1 million years ago).

According to plate tectonics, the landmasses of the Earth have been moving for millions of years. As shown in Figure 7.4, the landmasses have usually been separated by vast bodies of water. At one point in geologic history, all modern continents combined to form a single, great continent called Pangaea. The supercontinent Pangaea existed approximately 250 million years ago during the Permian Period. The existence of a single continent has a special significance in the context of the origin and evolution of life.

The solid crust of the Earth first formed approximately 4 billion years ago, followed shortly thereafter, by the Earth's atmosphere and oceans. Scientists have found traces of life in fossils older than 3.5 billion years. The existence of a single landmass, Pangaea, billions of years after the origin of life suggests that life could have originated at a single region on Earth before Pangaea formed, and then spread onto all of the continents without having to cross great bodies of water. Indeed, fossil evidence has been found showing the existence of such relatively advanced life forms as mammal-like reptiles and dinosaurs at the same time that Pangaea existed.

Pangaea did not retain its unity. Basalt extrusions from the mantle split the Pangaean plates. Gradually Pangaea began breaking up. Fragments of the Pangaean supercontinent continued moving and they now occupy their present configuration illustrated in Figure 7.4.

Personal Anecdote: Volcano National Park

One of the best ways to develop an appreciation of the scale of geologic forces is to view them up close and from a distance. Mount Kilauea on the "Big Island" of Hawaii is in Volcano National Park. Visitors to the park can see geologic forces at work. I had a chance to walk in a caldera, across lava fields, through lava tubes, and see new land being born. I used a picture of lava flowing into the Pacific Ocean as the cover picture for my book *Energy in the 21st Century* [Fanchi, 2005]. The picture was taken from the passenger seat of a helicopter and illustrated geothermal energy.

The Miller–Urey Experiments

One of the most important pieces of evidence[1,5] in support of Oparin's and Haldane's spontaneous generation hypothesis was provided in 1953 by American chemists Stanley L. Miller (1930–2007) and Harold C. Urey (1893–1981). Miller was a graduate student of Urey's when he performed an experiment of great significance to the development of a modern scientific theory of the origin of life. Miller used an apparatus similar to the one sketched in Figure 7.5. At the bottom left-hand side of the figure is a flask containing liquid water and three gases: hydrogen, ammonia, and methane. The fluids in the flask are a mixture known as the prebiotic soup. The prebiotic soup consisted of inanimate materials which were hypothesized to be components of the Earth's atmosphere prior to the appearance of life. All of the prebiotic constituents — water, hydrogen, methane, and ammonia — are relatively simple compounds. In addition to hydrogen atoms, water contains oxygen, ammonia contains nitrogen, and methane contains carbon. Gases such as methane and ammonia could have been expelled into the atmosphere of the primordial Earth by volcanic activity or by outgassing from the Earth's crust. Water was present in the Earth's oceans, and hydrogen existed in abundance throughout the universe.

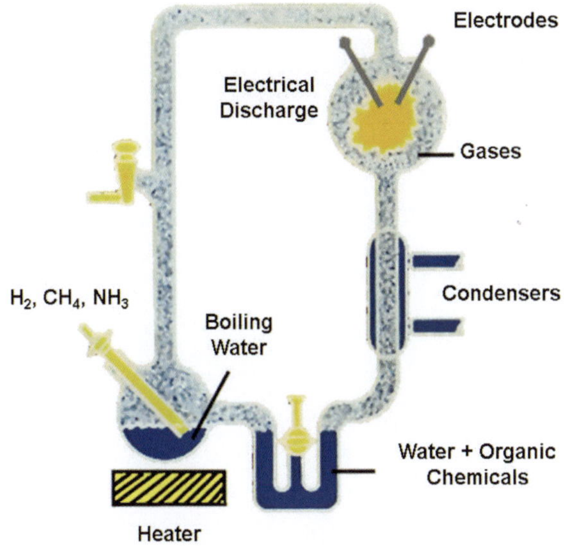

Figure 7.5. Sketch of the apparatus for the Miller–Urey experiments.

In addition to the prebiotic soup, ultraviolet radiation from the Sun, and volcanic activity were some major heat sources that were in contact with the Earth's atmosphere. These heat sources are represented by the heater under the flask containing the prebiotic soup. As the water boils, it vaporizes. The resulting steam mixes with the other gases and warms them. These heated gases are forced to flow through the path bounded by the glass tubes. Once the gases are thoroughly mixed, they are subjected to electrical discharges (sparks). Electrical discharges are supposed to represent energy sources such as lightning flashes, geothermal heat, ionizing radiation such as cosmic rays, or ultraviolet light irradiating the primordial Earth. The products of this activity are condensed in the water-cooled condenser and re-circulated.

After running the experiment for a week, the spark was turned off and the end products were analyzed. Among the chemical products were four naturally occurring amino acids: glutamic acid, glycine, alanine, and aspartic acid. Since the publication of the Miller–Urey results [Miller, 1953; Bada and Lazcano, 2012], other researchers have performed similar experiments. According to Rauchfuss [2008, p. 88], experiments like the Miller–Urey experiments "provided the first proof that the question of the origin of life is a scientific problem which can be approached (and possibly solved) by using scientific methodology." The Miller–Urey experiments marked "the beginning of prebiotic chemistry and modern biogenesis research" [Rauchfuss, 2008, p. 89]. Rauchfuss chose to use the word "biogenesis" to refer to origin of life research [*Ibid.*, 2008, p. 1].

The Oparin–Haldane–Miller–Urey model assumes the atmosphere of the primordial Earth was reducing, that is, the atmosphere was hydrogen-rich. Norman Horowitz observed that "Many geologists now doubt that a strongly reducing atmosphere ever existed on the Earth" [Horowitz, 1986, p. 41]. If the Earth's atmosphere was not reducing, and the approach by Miller and Urey is wrong, then where did life come from?

Besides producing important biochemicals from simple molecules under prebiotic conditions, people have learned through experimental tests that it is unlikely that amino acids can be produced from our current atmosphere composed of nitrogen (78%), oxygen (21%), and approximately 1% of other gases such as argon and carbon dioxide. This observation and the measurement of iron in the fossil record provide evidence that the atmosphere of the primordial Earth was different than it is today.[6]

The idea that the amount of free oxygen in the atmosphere depends on interactions between biological, environmental, and geological factors was championed by twentieth century American paleontologist Preston Cloud (1912–1991) [Cloud, 1983]. Development of an oxygen-rich atmosphere did not occur until oxygen-producing organisms like cyanobacteria had evolved. Cyanobacteria are prokaryotes that use photosynthesis to produce oxygen. Oxygen began to accumulate in the atmosphere about 2.5 billion years ago.

Attempts to synthesize the molecules of life from inorganic constituents on a primordial Earth have had some success. Sri Lankan chemist Cyril Ponnamperuma (1923–1994) [Ponnamperuma, 1974] performed experiments under primordial Earth conditions in which amino acids are produced and link together to form polymers. American biochemist Sydney W. Fox (1912–1998) [Fox, 1997] showed that these polymers acquire a spherical shape (called microspheres) when exposed to a watery environment similar to an ancient tidal pool. Earth scientist Robert Hazen (1948–) [Hazen, 2005] conducted work on the origin of life in the context of the science of emergence. The science of emergence is the study of complex systems which "display novel collective behaviors that arise from the interaction of many simple components" (p. xiv). According to Hazen, "life is the quintessential emergent phenomenon" (p. 13). Hazen's group was interested in studying the synthesis of organic molecules at conditions associated with hydrothermal vents.

Despite these successes, there are many gaps in our understanding of the evolutionary mechanisms leading from simple molecules to unicellular organisms. Scientists have been unable to synthesize even the simplest living organism in the laboratory. This has led some to consider yet another possible source of life on Earth: an extraterrestrial source.

Panspermia

The theory of Panspermia[7] hypothesizes that life was brought to Earth from outer space. It was introduced by Swedish chemist Svante Arrhenius (1859–1927) in 1908 because of the inability of science to explain the origin of life at the beginning of the twentieth century. By the middle of the twentieth century, people thought the Miller–Urey experiments provided a plausible mechanism for the origin of life, but questioned the validity of its assumptions about the atmosphere of the early Earth. Although the hypothesis of a reducing atmosphere assumed by Miller and Urey has not been disproved, other alternatives, like Panspermia, are being considered. Is there any evidence in support of Panspermia?

On September 28, 1969, a meteorite landed near the town of Murchison, Australia. Pieces of the meteorite were carefully collected and analyzed. The analysis found relatively large amounts of the same natural amino acids produced in the Miller–Urey experiments. This discovery, based as it was on very careful analyses designed to minimize the possibility of Earthly contamination, demonstrated that some of the building blocks of life exist elsewhere in the universe. The Murchison meteorite, as it is called, provided encouragement to proponents of the belief that life on Earth may have originated somewhere in outer space, and that extraterrestrial life exists.

Another rock found in the Antarctic was identified as a meteorite from Mars in 1996. This meteorite is known as ALH84001, where ALH refers to Allan Hills in Antarctica. Study of the rock revealed a mineralogy and structures that some researchers attributed to biological activity, presumably while still on Mars. Alternative explanations are possible that do not require such an extrapolation of normal science [Schulze-Makuch and Irwin, 2018, Chapter 10].

Even if extraterrestrial life exists, it does not prove life on Earth was brought here by a spore on a meteorite or by alien space travelers. Arguing that life on Earth originated somewhere in outer space does not solve the problem of the origin of life; it merely shifts it from Earth to outer space. Russian rocket scientist Konstantin Tsiolkovski (1857–1935) and Italian physicist Enrico Fermi (1901–1954) raised an obvious question that is now known as the Fermi paradox.[8] If sentient life is common, where is it? The question of the existence of extraterrestrial life is unresolved.

The Drake Equation

Frank Drake (1930–) of Cornell University developed an equation for computing the probability of the existence of extraterrestrial life [SETI-Drake, 2020]. The terms in his equation are conceptually easy to understand, but they require making estimates of quantities which are poorly known. Depending on your particular bias, you can use the Drake equation to show that extraterrestrial life is both likely and unlikely. Another Cornell University professor, astronomer Carl Sagan (1934–1996), used the Drake equation to show that extraterrestrial life is probable.[9]

Several authors have used the Drake equation or a variation to argue that life on Earth is rare. Peter D. Ward (1949–) and Donald E. Brownlee (1943–) [Ward and Brownlee, 2000] proposed the Rare Earth Hypothesis that the probability of finding microbial life elsewhere in the universe may be high, but the probability of finding more complex life forms will be low because many additional requirements must be satisfied for intelligent life to emerge. Paleontologist Andrew Knoll (1951–) [Knoll, 2003] said that Ward and Brownlee's argument that many conditions must be "just right" for intelligent life to exist is sometimes lampooned as the "Goldilocks hypothesis" after the little girl in the children's story "Goldilocks and the Three Bears." Teerikorpi *et al.* [2019, pp. 549–551] discussed the "just right" conditions as fine-tuning the physical and biological conditions that led to the development of a sentient species. The relationship between the existence of self-aware, sentient species, and cosmology has become part of mainstream cosmology as the Anthropic Principle, which we discuss in more detail later.

Alternative Biogenesis Paradigms

The standard paradigm of the origin of life uses chemical evolution as its basic paradigm. Chemical evolution is based on a few fundamental assumptions with far-reaching ramifications. It is believed that the most important biochemicals, the chemicals of life, are the same for all organisms on Earth today; that these chemicals were synthesized from simpler chemicals that existed on the primordial Earth; and that the chemicals of life must be carbon-based, that is, they must contain carbon. Is the experimental basis for the paradigm of chemical evolution sufficient to withstand challenges from competing paradigms?

Clay: The First Organism?

Chemist A. Graham Cairns-Smith (1931–2016) [Cairns-Smith, 1985a, 1985b] challenged each of the fundamental beliefs of the chemical evolution paradigm. He suggested that the synthesis of carbon-based chemicals from simpler carbon-based chemical building blocks is an idea that has not stood the test of time. The Miller–Urey experiments and experimental work since then have shown that many of the same biochemicals can be made under a wide range of conditions, not just those on the primordial Earth. Cairns-Smith argued that the unity of biochemistry does not apply to the start of life, but to a much later stage in the evolutionary process. The lowest common ancestor of all life on Earth today was not the first organism, but a descendent of other organisms that may have been considerably different. He agreed that evolution should arrive at the best chemical basis for life, which is probably carbon-based, but he argued that evolution should start with the easiest.

Cairns-Smith described a few characteristics a first organism should have. The foremost characteristic is the ability to evolve. The organism must contain material whose form is similar from one generation to another. This genetic material should serve as a template for future generations. The template must not be perfect, however. Evolution can only occur if changes are possible. These changes are mutations or defects in the genetic template. Mutations must be transmittable to, or inheritable by, subsequent generations. What should this genetic material be?

Cairns-Smith said the material should be made of commonly occurring geochemicals. These chemicals would be the easiest starting point of evolution, even though they may not be the best basis for life. In addition, the first organism should be simple relative to modern carbon-based organisms, but not so simple that a substantial amount of information cannot be contained in its structure. Consequently, the first organism must be made of many atoms. Cairns-Smith said that clay is a material that has all the qualities he identified.

Clay is a layered silicate, that is, it consists of layers of silicon-oxygen compounds. It is a commonly occurring geochemical. Clay may consist of as many atoms as the simplest carbon-based organisms alive today, which is enough atoms to carry genetic information. Clay forms easily by spontaneous crystallization. It has a reproducible structure which can be viewed as a template for future generations, yet the structure can acquire defects that are transmittable. If we accept clay or another silicon-based material as a legitimate contender for the title of "first organism," we must explain how life evolved from a silicon-base to a carbon-base.

Cairns-Smith suggested that genetic takeover is the mechanism by which modern life evolved from clay. Genetic takeover is the replacement of one gene material (such as silicon-based clay) with another (such as carbon-based chemicals) over a series of generations. Cairns-Smith viewed clay as an incubator within which chemical evolution could occur. The clay incubator was discarded when chemical evolution had progressed to the point where the clay incubator was no longer needed.

Deep, Hot Biosphere

The mainstream theory of the origin of oil and gas is called the biogenic theory. It says that oil and gas formed as a result of the death, burial, decay, and compaction of organisms that lived on or near the surface of the Earth. An alternative hypothesis called the abiogenic theory, says that processes deep inside the Earth, in the Earth's mantle, form petroleum. Austrian astrophysicist Thomas Gold (1920–2004), an advocate for the abiogenic theory [Gold, 1999], pointed out that the biogenic theory was adopted in the 1870s. At the time, scientists thought that the Earth was formed from molten rock that was originally part of the Sun. In 1846, Lord Kelvin, also known as William Thomson (1824–1907), estimated the age of the Earth from the rate of cooling of molten rock to be about 100 million years old. A more accurate estimate of the age of the Earth was made after French physicist Antoine Henri Becquerel (1852–1908) discovered radioactivity in 1896. In 1905, New Zealand chemist Ernest Rutherford (1871–1937) proposed using radioactivity to measure the age of the Earth from the concentration of long-lived radioactive materials in rock. The Earth is now believed to be approximately 4.5 billion years old.

Gold's abiogenic theory is consistent with the Oparin-Haldane hypothesis that life emerged from inanimate material under the prevailing conditions of the primordial Earth. He argued that simple inorganic and organic molecules in the forming Earth were subjected to increasing heat and pressure, and eventually formed more complex molecules. These complex molecules eventually became the simplest forms of life.

Gold presented several pieces of evidence in support of the abiogenic theory and the possible existence of a biological community deep inside the Earth. Some of Gold's evidence included the existence of microbial populations that can thrive in extreme heat. These microbes, notably bacteria and archaea, grow at hot, deep ocean vents and can feed on hydrogen, hydrogen sulfide, and methane. Hot, deep ocean vents are examples of extremophilic environments.

Gold considered life forms at deep ocean vents transitional life forms that exist at the interface between two biospheres. One biosphere is the surface biosphere and includes life that lives on the continents and in the seas on the crust of the Earth. Gold postulated that a second biosphere existed in the mantle of the Earth that he called the deep biosphere. The surface biosphere uses chemical energy extracted from solar energy, while the deep biosphere feeds directly on chemical energy. Oxygen is a requirement in both biospheres. Gold's deep biosphere was the source of life that eventually formed hydrocarbon mixtures (petroleum) in the Earth's mantle. Crustal oil and gas reservoirs are formed by the upward migration of petroleum fluid until the fluid is stopped by impermeable barriers and accumulates in geological traps.

Personal Anecdote: Anaerobic Bacteria

The abiogenic theory depends on the existence of organisms that do not require an oxygen-bearing environment to live. I first encountered anaerobic bacteria in 1979 when I was modeling the injection of expensive fluids into an oil reservoir to improve the recovery of oil. One of the fluids was a polysaccharide, a high molecular weight polymer that was supposed to increase the viscosity of injected water. Instead, it appeared that the polysaccharide was disappearing. A biologist and a chemist studied the problem and discovered that the polysaccharide was a source of food for anaerobic bacteria. The little critters were eating the expensive polymer.

Extremophilic Origin

Cairns-Smith's theory is an example of a silicon-based paradigm. Gold's deep, hot biosphere is an example of an anaerobic environment. Other theories have been proposed, but research is continuing.[10] We have already introduced examples of extremophilic environments where extreme conditions of temperature, pressure, acidity, salinity, or other challenging chemical environments exist. Extremophiles are carbon-based organisms that live in extremophilic environments. These conditions are unsuitable for most life on Earth. Extremophiles often depend on chemosynthesis rather than photosynthesis. Chemosynthesis uses the energy from chemical reactions, rather than light, as the source of energy for biologically converting molecules with a single carbon atom such as carbon dioxide or methane, into organic matter. Archaea are an example of a type of extremophilic organism. Archaea are thermophilic, single-celled organisms without a nucleus, so archaea are prokaryotes. Both eukaryotes and prokaryotes are part of the life transition summary presented in Table 6.1.

German chemist Günter Wächtershäuser (1938–) argued that the first organisms originated in or near an extremophilic environment such as a submarine hydrothermal boundary at a divergent plate boundary or hotspot.

Figure 7.6. Black smoker [2021].

A geological example is a black smoker, that is, a vent of hot, mineralized water formed in mid-ocean fracture zones. A sketch of a black smoker is shown in Figure 7.6.

Wächtershäuser proposed the iron-sulfur world hypothesis that said early life may have formed on the surface of minerals such as iron sulfide (pyrite) in the vicinity of deep (volcanic), hydrothermal vents. Ocean water can flow through fractures into porous formations that are heated by a heat source such as a magma chamber. The heater water can carry dissolved materials with it through tubes in the vent until it returns to the ocean floor. Bubbles exiting the vent act as "cells" to localize materials and facilitate reactions. The movement of heated water to a cooler environment transports produced polymers to a lower temperature environment where cooler water prevents heat from destroying the polymers.

In this scenario, heated water and minerals help synthesize polymers of amino acids. Amino acids are formed by mixing inorganic chemicals (carbon monoxide, hydrogen sulfide, nickel sulfide, and iron sulfide) from mineral deposits in the hot, high-pressure environment. Wächtershäuser [2006] called the first organism the pioneer organism.

❁ ❁ ❁

Origin of life researchers are trying to find the correct path to the selection, concentration, and organization of prebiotic organic molecules into structures that exhibit the essential traits of life: the ability to grow, reproduce, and evolve. According to Rauchfuss [2008, p. 87], biomolecules associated with the origin of the first living organisms could have originated from one of the following three possibilities:

1. They could have been synthesized from inorganic molecules in the atmosphere, hydrosphere, or lithosphere of the early Earth.
2. They could have been brought to Earth from outer space by meteorites or comets.
3. They could have originated from a combination of origins 1 and 2 above.

Rauchfuss [2008, p. 87] concluded that "we still do not know which of these three will lead us to the most convincing answers to the many questions which still remain unanswered." Schulze-Makuch and Irwin [2018, p. 50] agreed that we have "huge gaps in our knowledge of how life actually began on Earth (or an alien world prior to its transport to Earth)." Nevertheless, they supported the idea that we can postulate a plausible scenario for developing living organisms from prebiotic materials.

Following a review of the origin of life research, McKay [2020, p. 195] observed that while evidence for the origin of organic materials relevant to biochemistry from nonbiological processes "is compelling, the standard theory for the origin of life suffers from a major flaw — it has not been possible to create life" from nonbiological processes. He said that new data were needed to solve the problem of the origin of life, and concluded that "even the common assumption that life originated on Earth is not directly supported, or refuted, by data" [McKay, 2020, p. 196].

The intellectual step from the results of laboratory experiments to the conclusion that life can be constructed from inanimate materials or nonbiological processes is a leap of faith. It is one leap in a series of leaps that must be made to explain the evolution of people. In the next chapter, we discuss how more complex organisms have evolved.

Chapter 8

The Evolution of Life

Scientists would like to develop a theory of the evolution of living organisms, from the simplest to the most complex, using only materials present in the physical universe. The origin of unicellular organisms from nonliving material was discussed in the previous chapter. Here we examine the mainstream scientific view of the evolution of complex life from unicellular organisms.[1,2]

The unifying theory of biology is the theory of evolution. It was first formally proposed by Englishman Charles Robert Darwin (1809–1882) and codiscoverer Alfred Russel Wallace (1823–1913) in 1858. In this chapter, we outline the historical context of Darwinian evolution by considering some pre-Darwinian perspectives on evolutionary concepts. We then discuss the Darwin-Wallace theory of evolution and its successor, the synthetic theory of evolution.

Pre-Darwinian Evolution

The dawn of evolutionary thought can be traced to the early Greeks. Anaximander (ca. 610–546 BCE) looked for natural rather than supernatural explanations of the origins of living things. Aristotle (384–322 BCE) accepted the primacy of sense experience, and believed that reason is secondary. Aristotle introduced the Scala Naturae or ladder of life. He believed that organisms rose on the ladder of life by increasing perfection.

French naturalist Georges-Louis Leclerc de Buffon (1707–1788) introduced ideas of geographic variation and change but did not specifically discuss evolution.

James Hutton of Scotland (1726–1797) introduced the concept of uniformitarianism, that is, the idea that the present is the key to the past. He believed that Ussher's estimate that the Earth is 6,000 years old was too young to account for geological observations such as the gradual formation of canyons. Hutton argued that the earth must be very old to allow geologic events to occur gradually.

Hutton and Charles Lyell (1797–1875) of Scotland were proponents of gradualism and uniformitarianism. Gradualism is the belief that large-scale natural change or variation occurs gradually. The observation of marine fossils in mountains such as the Andes supported Lyell's view that the earth had to be very old to allow time for marine fossils to move from the sea floor to a mountaintop. In addition, Lyell wrote *Principles of Geology*, which consisted of three volumes published between 1830 and 1833, that popularized the notions of gradualism and uniformitarianism. Charles Darwin took Lyell's *Principles of Geology* as a reference on his voyage aboard the H.M.S. Beagle.

Jean-Baptiste Lamarck (1744–1829) of France subscribed to Aristotle's ladder of life. Species on the upper rungs of the ladder of life were thought to have achieved more perfection in life. How did species move up the ladder of life? Lamarck suggested that offspring inherit characteristics acquired during life. He believed that features acquired by parents are passed on to offspring. As an illustration, suppose giraffes acquired long necks to reach leaves. Long necks would be considered an acquired characteristic that was passed on to their offspring. Today it is thought that acquired characteristics are not inheritable, for example, the loss of limbs by a parent is not passed on to the offspring.

The Darwin — Wallace Theory of Evolution

Charles Darwin was the son of an affluent physician and grandson of Erasmus Darwin (1731–1802). Erasmus Darwin was a physician and natural philosopher who explicitly discussed evolution. For example, Erasmus wrote about the descent of life from a common ancestor in Volume I of *Zoonomia* [Darwin, 1794, pp. 545–546]. He wrote that "all animals have a similar origin, viz. from a single living filament; and that the difference of their forms and qualities has arisen only from different irritabilities and sensibilities, or voluntarities, or associabilites, of this original living filament … the great variety of species of animals, which now tenant the earth, may have had their origin from the mixture of a few natural orders." Erasmus' work influenced his grandson's thinking about evolution.

Charles enrolled in Edinburgh University, Scotland in 1825 to study medicine, but found that he was unable to watch a surgery. He entered Christ's College in Cambridge, England in 1828 to study for the clergy. Instead, Charles developed an interest in natural history and received a bachelor's degree from Cambridge in 1831.

Darwin accumulated observational evidence for his theory of evolution while serving as naturalist on board the H.M.S. Beagle. The H.M.S. Beagle under the command of Captain Robert Fitzroy (1805–1865) departed England in December, 1831 on a 5-year around-the-world voyage. Most of the voyage was spent surveying the South American coast. When the position of ship's naturalist aboard the H.M.S. Beagle became available, one of Charles Darwin's instructors recommended 22-year-old Charles for the position.

Charles Darwin had many experiences on the voyage that influenced his understanding of life. He studied coastal fossil beds and collected rain forest specimens. He observed human slavery in Argentina, experienced an earthquake in a coastal Chilean town, and watched the shore rise more than a foot above sea level in seconds. His experience with earthquakes suggested a mechanism that could explain the presence of fossil seashells he collected at high altitudes in the Andes Mountains.

The H.M.S. Beagle spent time at the Galapagos Islands west of South America near the end of its voyage. Darwin studied giant tortoises with features that were unique to each island. He observed species of finches that were adapted to each of their habitats. For example, finches on dry islands developed stout beaks and fed on cracked seeds. By contrast, finches on wetter islands developed smaller beaks and fed on fruit or insects. How could such diversity arise?

During his transit time on the voyage, Darwin had the opportunity to read Charles Lyell's (1797–1875) *Principles of Geology*. He developed a sense of deep time in his study of geology and acquired a better understanding of fossils in relation to such geologic events as earthquakes. He observed the variability of species in such species as tortoises and finches. How did geographic separation on the islands foster the development of new species, which is now known as speciation? He noticed the stability of natural populations: common species were the dominant species within their natural range while rare species were rare in that range.

Darwin returned to England in 1837. In 1838, he read an essay by Thomas Malthus (1766–1834) on the size and growth of human population. Malthus was concerned that human population was rapidly increasing and could lead to starvation, disease, and death as the result of an inadequate food supply. He hypothesized that human societies were in competition for scarce resources. Darwin realized that there was a difference between reproductive potential and reproductive success.

Darwin used the term "natural selection" to describe the process by which populations are unable to reach their maximum reproductive potential because less fit individuals perish while the fittest individuals survive. Darwin defined the fittest individuals as those individuals that are best adapted to survive in an environment subject to limited natural resources, disease, catastrophic events, etc. He adopted the phrase "survival of the fittest," which was a phrase coined by English philosopher Herbert Spencer (1820–1903) in his 1864 book *Principles of Biology*. Today we consider the fittest those individuals which survive to reproductive age and contribute to the gene pool.

Darwin applied the concept of competition for scarce resources to explain the evolution of species. Those species with an unusual characteristic, such as great speed or strength, or intellect, would survive and reproduce. Other species, being at a disadvantage, would be unable to compete for essential life-sustaining resources and would eventually become extinct. Thus, small herbivores that were swift of foot or clever had a better chance of surviving than comparably sized herbivores who could not outrun or outsmart predators.

Darwin published his theory when he learned that fellow Englishman, naturalist Alfred Russel Wallace, had independently developed the same theory and was ready to present his work. Wallace did much of his work in the Malay Archipelago, Indonesia. Both Darwin and Wallace were influenced by a familiarity with the social philosophy of economist Thomas R. Malthus and philosopher Herbert Spencer (1820–1903). Wallace's concepts were essentially the same as Darwin's, including the view that evolution is a struggle for survival, but not supported by as much observational data.[2]

According to A.C. Brackman in *A Delicate Arrangement: The Strange Case of Charles Darwin and Alfred Russel Wallace* [Brackman, 1980], Darwin had accumulated more documentation than Wallace, worked on the theory longer, and had better professional connections, for example, geologist Lyell and botanist Joseph D. Hooker (1817–1911). After a series of discussions with leading members of the Linnean Society, Darwin was assigned primacy and Wallace was assigned co-discover of the Darwin-Wallace theory of evolution. The theory was submitted as a joint presentation to the Linnean Society of England on 1 July 1858.

Darwin's work with Wallace prompted Darwin to publish *On the Origin of Species* on 24 November 1859. Today, Darwinian evolution, or Darwinism, refers to Darwin's published work. The success of Darwinian evolution raised questions about the intellectual predecessors who contributed to the theory of evolution. Darwin added a new section entitled "An Historical Sketch of the Recent Progress of Opinion on the Origin of Species" to the 4th edition of *On the Origin of Species by Natural Selection* to address these questions [see Stott, 2017, Appendix].

Mayr's Summary of Darwinian Evolution

Most naturalists before Charles Darwin believed that species were immutable, that is, they did not change. This belief was one of the beliefs that changed following the publication of Darwin's theory of evolution. Species were no longer considered "immutable." German-born American biologist Ernst Mayr (1904–2005) succinctly summarized Darwinian evolution in terms of five facts and three inferences [Mayr, 1982, pp. 479–480]. Facts 1 through 3 from population ecology are presented in Table 8.1 and led to Inference 1

Table 8.1 Mayr's facts.

Fact	Facts from Population Ecology
1	"All species have such great potential fertility that their population size would increase exponentially … if all individuals that are born would again reproduce successfully."
2	"Except for minor annual fluctuations and occasional major fluctuations, populations normally display stability."
3	"Natural resources are limited. In a stable environment they remain relatively constant."
	Facts from Genetics
4	"No two individuals are exactly the same; rather, every population displays enormous variability."
5	"Much of this variation is heritable."

Table 8.2 Mayr's inferences.

Fact	Inferences
1	"Since more individuals are produced than can be supported by the available resources but population size remains stable, it means that there must be a fierce struggle for existence among the individuals of a population, resulting in the survival of only a part, often a very small part, of the progeny of each generation."
2	"Survival in the struggle for existence is not random but depends in part on the hereditary constitution of the surviving individuals. This unequal survival constitutes a process of natural selection."
3	"Over the generations this process of natural selection will lead to a continuing gradual change of populations, that is, to evolution and to the production of new species."

in Table 8.2. Combining the facts from population ecology with Facts 4 and 5 from genetics in Table 8.1 led to Inferences 2 and 3 in Table 8.2.

The Synthetic Theory of Evolution

Darwinian evolution supported the notion that species evolve gradually and continuously, and disputed the theory of catastrophism pioneered by French biologist Baron Georges Cuvier (1769–1832). Cuvier was a founder of the science of paleontology, the study of fossils. He demonstrated that the fossil record could be used to yield meaningful information about extinct animals. He believed that fossils in a particular rock stratum could be used as an identifying characteristic of that stratum. Cuvier explained the abrupt differences in fossil content from one stratum to the next in terms of widespread environmental catastrophes. These catastrophes were supposed to have destroyed all animals in a given geologic stratum. Life in subsequent geologic strata resulted from new creations. In this way, Cuvier could also explain the biblical Noah's flood.

Cuvier's catastrophism seemed to work well for the observations he had, but his detailed observations were limited to the Paris vicinity. English geologist William Smith (1769–1839) made detailed observations throughout England. Analysis of Smith's work showed that evidence for Cuvier's catastrophes in the fossil record could be found in many strata, but not everywhere. It became clear that Cuvier's catastrophes were neither worldwide nor even continent wide. Something was wrong with Cuvier's catastrophism.

An alternative explanation was provided by geologists James Hutton (1726–1797) and Charles Lyell in the 18th and 19th centuries. Hutton and Lyell argued that the present is the key to the past, that is, only contemporary processes — wind, rain, erosion, volcanoes, and so on — should be used to understand the development of the Earth over geologic time. These processes work slowly but continuously over long periods of time.

Darwin's understanding of the biological observations he accumulated during his tenure as naturalist on the H.M.S. Beagle was influenced by the geologic concepts he read in Lyell's *Principles of Geology* [Lyell, 1830–1833]. To explain the great variety of species he observed, Darwin postulated that the characteristics of species were always changing.

One possible cause of the changes Darwin noted was attributed to the external environment. This idea was first advocated by French naturalist Georges Buffon, and developed further by Buffon's countryman Jean-Baptiste Lamarck (1744–1829). According to Lamarck, parents in a given species must either adapt to the changing environment or eventually become extinct. For those species which adapt, their new characteristics are passed on to their offspring by heredity. Lamarck's ideas have been superseded by mutation theory, which is based on the genetic theory of Austrian priest and naturalist Gregor Johann Mendel (1822–1884).

Mendel studied the edible pea in a monastery garden at Brunn, Austria. By crossing pea plants, he was able to produce obvious differences in characteristics, such as changes in the size of the plant and the color of its flowers. Mendel observed that the traits of the offspring depended on the traits of the parents and even earlier generations. He realized that it was not the characteristics that were inherited, but the factors determining the characteristics. Mendel's work led to the development of the concept of genes as the carriers of information about the characteristics of an organism. Genes are not altered by structural adaptations of the parent. Today it is known that hereditary characteristics are conveyed from one generation to another through genes in cells. Changes in genes, called mutations, provide a mechanism for explaining how variations within a population can occur.

According to Darwinian evolution, the variety of life we see on Earth today descended from a common ancestor. The changes in the lineage of a species from one generation to the next were attributed to natural selection, or survival of the fittest. One of the main scientific difficulties with Darwinian evolution was the ambiguity associated with the concept of natural selection. The combination of descent with modification from a common ancestor, and Mendel's genetics became the basis for a new theory of evolution called neo-Darwinian evolution. In neo-Darwinian evolution, life progresses from simple to complex and evolution by natural selection is driven by successful mutations. Species diversify as time moves from the past to the present.

The discovery of more fossils in geologic formations like the Burgess shale led scientists to question the hypothesis that the evolution of life is a progression.[3] Many life forms found in the Burgess shale have not survived and do not have any surviving descendants. An unsuccessful lineage could be decimated and disappear during the course of geologic history. Figure 8.1 (a) shows diversification, while both diversification and decimation are shown in Figure 8.1 (b). Furthermore, there is no obvious trend toward greater complexity and diversity. Instead, it appears that evolution is driven by random variation and away from a simple beginning. A new view of evolution was needed.

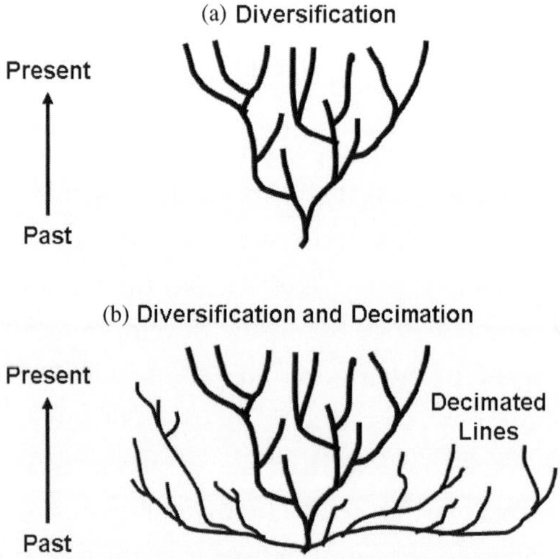

Figure 8.1. Diversification and decimation.

The modern view of evolution is the modern synthetic theory of evolution. In this theory, genetic variation occurs within a population as a result of random mutation or recombination of genes. Populations evolve when there is a change in the frequency of occurrence of a gene. The changes can be due to random genetic drift or natural selection. In the original version of the modern synthetic theory of evolution, the genetic variations occurred gradually and accounted for the observed diversification of species. The emergence of a new species by the process of genetic variation followed by natural selection is called speciation.

There are two schools of thought regarding the rate of evolution. The original school of thought is referred to as Gradualism. Gradualists view evolution as a gradual, continuous process. The new school of thought called Punctuated Equilibrium was introduced in 1972 by Americans Niles Eldredge (1943–) and Stephen Jay Gould (1941–2002). Proponents of Punctuated Equilibrium believe that evolution occurred rapidly over geologic time. The periods of rapid change were followed by periods of stasis, or relative calm. Thus, periods of evolutionary equilibrium were disrupted, or punctuated, by periods of rapid change.

Extinction events are examples of periods of rapid change.[4] A mass extinction is the rapid extinction of a large fraction of species. Two examples of extinction events that provide evidence to support the hypothesis of punctuated equilibrium are the Permian-Triassic extinction and the Cretaceous-Paleogene extinction. The Permian-Triassic mass extinction is characterized by the disappearance of trilobites from the fossil record at the end of the Permian Period, and the Cretaceous-Paleogene mass extinction is characterized by the disappearance of dinosaurs from the fossil record at the end of the Cretaceous Period. Geologic periods are listed in Table 7.1.

The Human Organism

For centuries biologists have sought to develop schemes for classifying living organisms of all kinds. The first major classification scheme, or taxonomy, was developed by Swedish naturalist Carl Linnaeus (1707–1778) and published as *Systema Naturae* in the twentieth century.[5] His scheme, like most taxonomies, used physical characteristics to distinguish one organism from another. It categorized organisms into one of two kingdoms: Plant, and

Animal. Linnaeus' two kingdoms were the standard until the 1960s. Advances in our knowledge of evolution and of the variety of organisms, especially microorganisms, led researchers to suggest new classification schemes.

American ecologist Robert H. Whittaker (1920–1980) replaced the two-kingdom approach with five kingdoms. His five kingdoms are the Monera, Protista (or Protoctista), Fungi, Plantae, and Animalia. Monera are unicellular prokaryotic organisms such as bacteria and cyanobacteria. Prokaryotes have a single strand of DNA, no nucleus, and are asexual. Protista are unicellular eukaryotic organisms, such as protozoa, single-celled fungi, and algae. Unlike prokaryotes, eukaryotes have nuclei; their DNA exists as a double helix with two interwoven strands; and reproduction is sexual. Sexual reproduction allows the mixing of genetic information from two parents. Whittaker's Kingdom Fungi are the multicellular fungi. Kingdoms Plantae and Animalia represent multicellular plants and animals respectively. They include organisms known to Linnaeus and his contemporaries.

American geneticist Carl R. Woese (1928–2012) introduced three domains: Archaea, Bacteria, and Eukarya (Figure 8.2). Each domain is a separate evolutionary lineage [OSBio, 2018, pp. 25–26]. Membership in a domain depends on genomes, where a genome is the complete set of genes and noncoding sequences of DNA that represent the hereditary history of an organism.

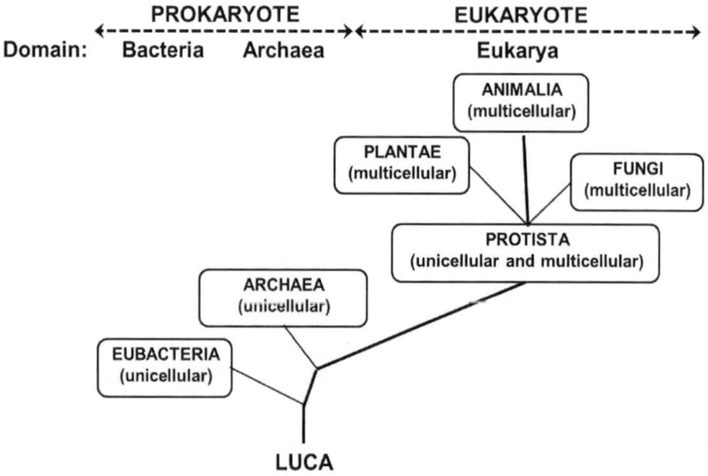

Figure 8.2. Domains and kingdoms.

Woese's three domains encompass Whittaker's five kingdoms. Whittaker's Prokaryotes were all unicellular organisms without nuclei. Woese subdivided Whittaker's Prokaryotes into the domains Bacteria and Archaea because of genome differences. Archaea may be an ancient intermediate group between bacteria and eukaryotes because their molecular biology is more similar to eukaryotes than bacteria. Many Archaea are extremophiles: they are found in what are considered to be extreme environmental conditions, such as geysers, ocean bottom vents, oil wells, and cold habitats.

Whittaker's remaining four kingdoms (Animalia, Plantae, Fungi, and Protista) were combined into domain Eukarya because of genome similarities. LUCA in Figure 8.2 is the Last Universal Common (or Cellular) Ancestor. It is the most recent organism from which all other organisms on Earth have a common descent.

Human Beings

Taxonomic classification gives us another perspective on our niche in nature. One set of taxonomic categories includes Domain, Kingdom, Phylum, Class, Order, Family, Genus, and Species. According to this taxonomic scheme, our place in nature is summarized in Table 8.3.

We are members of the domain Eukarya because our genetic material is contained in a distinct nucleus. We are part of Kingdom Animalia because we are multicellular eukaryotes that depend on other organisms for nutrition. Our embryonic development includes a stage during which we have gill pouches on our throats. Though the gill pouches disappear before birth,

Table 8.3 Taxonomic classification of Homo sapiens.

Domain	Eukarya
Kingdom	Animalia
Phylum	Chordata
Class	Mammalia
Order	Primates
Family	Hominidae
Genus	Homo
Species	Sapiens

their presence in our development, together with our skeletal and nervous systems, are used to classify us as members of Phylum Chordata. Our two arms and two legs place us in the Tetrapoda (four appendages) Superclass. The existence of body hair and mammary glands in the skin associate us with Class Mammalia. Our good vision and our ability to grasp with our hands place us in the Order Primates. An erect posture, relatively long arms, and no tail are the additional features exhibited by our family, the Hominidae Family. We share this family with the apes. Distinguishing us from the apes is an enlarged cerebral cortex, and a unique social behavior. Human intelligence is needed for survival within a complex social community and provides the basis for self-awareness, a uniquely human capability. These are the hallmarks of our genus: Homo.

Point to Ponder: What Is Cladistics?

The taxonomic scheme shown in Table 8.3 is based on shared characteristics. A more modern, and more detailed, classification scheme known as cladistics classifies organisms based on external features preserved in the fossil record and available genetic information. Organisms are grouped by clades, which are "groupings of organisms united by a shared and unique ancestor at some point in their evolutionary history" [Palmer, 2009, p. 250]. The history of life is represented by cladograms or nested clades.

The cladistic view of our place in the tree of life has been presented by Palmer [2009, pp. 250–295]. It provides a snapshot of our knowledge and understanding. As with many of the scientific topics covered in Part I, the cladistic view is subject to change when more observations are made and disputes are resolved.

Homo first appeared 2 to 3 million years ago[6] in the form of Homo habilis, or "handy man." The name was proposed because of an apparent connection to stone tools. It is possible that an older species may have existed such as Homo gautengensis. The genus Homo includes the Java and Peking men in Asia and the Neanderthal man in Europe. Our species, sapiens, appears to have evolved in Africa up to two hundred thousand years ago. Sapiens is a term meaning "wise." It is attached to our species because our cranial development is more pronounced than other species. Many other traits are needed besides brain size to classify a fossil as a modern human. They include shape of the skull, size of teeth and nose, and general body shape. Anatomically modern humans are known as Homo sapiens.

Point to Ponder: Does the Soul Have a Role?

Some scientists have suggested that modern humans have one more trait: the presence of a soul. Gerald Schroeder (1938–) makes the case that the Biblical Adam "may have been the first hominid with a divinely created human soul" [Schroeder, 1997, p. 126]. Francis S. Collins (1950–) commented that Biblical texts "seem to suggest that there were other humans present at the same time that Adam and Eve were expelled from the Garden of Eden" [Collins, 2006, p. 207]. For example, in Genesis 4:16–17, the *Bible* says that Cain found a wife after he left Eden. Collins concluded that "The real dilemma for the believer comes down to whether Genesis 2 is describing a special act of miraculous creation that applied to a historic couple [Adam and Eve], making them biologically different from all other creatures that had walked the earth, or whether this is a poetic and powerful allegory of God's plan for the entrance of the spiritual nature (the soul) and the Moral Law into humanity" [Collins, 2006, p. 207].

Alternatives to the Modern Synthetic Theory of Evolution

Syntheses from competing concepts have often led to significant advances in our understanding of the universe. One such synthesis was the combination of Darwinian natural selection and Mendelian genetics to form the synthetic theory of evolution. This theory has come under attack from two directions: creationists, and punctuationalists.

Creationism

Creationism covers a wide range of perspectives.[7] Examples of creationist thought include young-earth creationism, old-earth creationism, day-age creationism, and theistic evolution. Young-earth creationists believe the universe has an age ranging from 6,000 years to 10,000 years or so. Probably the most fundamental assumption of young-earth creationists is acceptance of Scripture as the authoritative word of God. Their beliefs are based on deductive arguments from Scripture, rather than inductive arguments from observations. Old-earth creationists accept the evidence for an Earth that is billions of years old but believe life was created by the biblical God. Day-age creationists interpret the Genesis day as a long period of time and see a relationship between events presented in Genesis 1 and the mainstream scientific paradigm. Theistic evolutionists believe that God used evolution to create man. For example, Pierre Teilhard de Chardin argued that God unified himself with nature by "partially immersing himself in things, by becoming 'element', and then, from this point of vantage in the heart of matter, assuming the control and leadership of what we now call evolution" [de Chardin, 1975, p. 294].

Some modern creationists have developed arguments to refute important aspects of the synthetic theory of evolution. A creationist argument against the synthetic theory is that geologic strata and fossils are dated in terms of each other, rather than by an absolute time scale. Evolutionists say that this argument betrays a lack of understanding of the methods of biostratigraphy. They point to radiometric dating as an example of an absolute time scale. Evolutionists are correct in this instance if the laws of physics, particularly radioactive decay mechanisms, have worked in the past as they do today. Cosmological inflation is an example of a theory where the rate of a physical process (expansion of the universe) was different in the past than it is today. If we note that radioactivity was not discovered until 1896 by French physicist Antoine Henri Becquerel (1852–1908), it is a considerable extrapolation to say that a process known for less than two centuries has applied unchanged for up to 15 billion years.

Another creationist argument says that the statistical probability of life arising from inorganic matter and evolving into complex life forms is very low. This argument is the basis of the Anthropic Principle, namely, that cosmology must allow for human existence. The creationist argument says that the probability that life could evolve from inorganic matter to its existing complexity implies that life as we know it must have been created and influenced by a supernatural entity to arrive at such an improbable result. Evolutionists reply that a statistically low probability is not impossible. They remind creationists that evolution is dealing with billions of years and a virtually endless supply of molecules. A point that is often overlooked in the debate is that evolution does not preclude the influence or existence of a supernatural entity.

One of the most important creationist arguments is that the fossil record does not contain transitional forms between different kinds of organisms. The operative word here is "kind." "Kind" is not the same as "species." Creationists recognize the fact of variation within "kind" that stems from environmental adaptation. They have little choice but to concede this point because adaptation has been demonstrated or observed in many instances, such as Mendel's peas and the variation of Galapagos Island finches noted by Darwin during his voyage on the H.M.S. Beagle. Evolution from one "kind" to another is rejected by creationists. Evolutionists say that the creationist argument is wrong because the creationist definition of "kind" is not precisely defined, and therefore scientifically untestable. The evolutionist rebuttal does not address the real issue raised by the creationists: where are the transitional forms in the fossil record? The search for transitional forms is being pursued by proponents of different schools of thought within normal science.

Punctuationalism

Proponents of the synthetic theory of evolution subscribe to a few basic evolutionary mechanisms: organisms vary within a population; the variation can be inherited; more offspring are produced each generation than can survive; and the organisms best suited to their environment tend to produce more viable offspring. Evolution is viewed as a gradual, adaptive process called anagenesis. According to this view, transitional forms linking one major taxonomic unit (or taxon) with another must have existed. Why, then, is there a limited set of examples of transitional forms in the fossil record? This scarcity is a legitimate scientific question for the synthetic theory. It has motivated the development of alternative approaches to evolution by anagenesis. Many scientists believe that punctuated equilibrium is an alternative that is more consistent with the data.

Punctuationalists, or proponents of punctuated equilibrium such as Eldredge and Gould, suggest that the fossil record supports an evolutionary process in which stasis is the norm and morphological change is the exception. Morphological change refers to the change in shape, structure, arrangement, and function of parts of organisms as a result of evolution. When morphological change does occur, it is rapid and usually results in speciation, the creation of new species by the branching or fission of lineages. The lineages are organized into clades, and the process of branching is known as cladogenesis. From a punctuationalist perspective, it can be argued that cladogenesis occurs when species are forced to adapt to a new environment.

❁ ❁ ❁

A popular conception about biological evolution is the inevitability of the evolutionary process: given the proper environment and the right chemical ingredients, life should evolve much as it has on Earth. This concept is often the basis of searches for extraterrestrial life. Is this a valid concept?

The theory of the evolution of life presented thus far is based on empirical observations. Though the observations are firmly established, their interpretation is not. It is the interpretation of observations and the intellectual leaps across gaps in the observational evidence that are a principal source of controversy between competing worldviews.

American geneticist Francis S. Collins, a Christian and long-time head of the Human Genome Project, observed that "The study of genomes leads inexorably to the conclusion that humans share a common ancestor with other living things" [Collins, 2006, p. 133]. It does appear to be true that natural selection favors individuals with variations that make them better able to exploit their environment. Beneficial genetic variations enhance the survivability of an individual and can be inherited by its offspring. Collins concluded that "Evolution, as a mechanism, can be and must be true. But that says nothing about the nature of its author" [*Ibid.*, p. 107].

We have now established a scientific view of the origin and evolution of the universe and life. The scientific worldview is effectively equivalent to a secular worldview. It is an argument that god is not necessary to explain the existence of the universe and life. Alternatives to the secular worldview presented in Part 1 are discussed in Part 2, beginning with mythological creation stories.

Part 2

The Ultimate Gamble

Chapter 9

Mythological Creation Stories

Part 1 is a summary of the scientific worldview. Our personal worldviews may or may not agree with the scientific worldview. Does that mean we are wrong? Should we let learned authorities, either scientific or religious, tell us their answers and then accept them without question? Some issues are so important that we should seek our own answers, especially since the views of science have a history of changing.

Part 2 introduces other worldviews that are representative of religious and philosophical perspectives that have been historically prevalent around the world. The conflict between worldviews results in a struggle to decide what we should believe, and what we should teach future generations. The struggle is often not between science and religion, however, but between reason, faith, and purpose. In this chapter, we consider creation myths from around the world.

Creation Myths

A myth is a popular belief or story that attempts to explain natural occurrences or the history of a people. Myths often associate supernatural entities with historical events, such as volcanic activity, floods, and earthquakes. The common usage of the word myth often suggests that any story labeled a myth is false. We are not implying here that the stories presented in this chapter are false. Our goal in this chapter is to illustrate the historical and worldwide interest in the origin of the cosmos by presenting a sampling of creation myths. These creation myths illustrate the complexity of thought underlying ancient mythologies.

Point to Ponder: Is Modern Science a Myth?

Old Testament scholar John H. Walton (1952–) observed that mythology "seeks to explain how the world works and how it came to work that way, and therefore includes a culture's 'theory of origins'" [Walton, 2009, p. 14]. He pointed out that the literature we now call mythology was once "a real description of deep beliefs" for some people [*Ibid.*, p. 14]. According to this definition of mythology, "our modern mythology is represented by science" [*Ibid.*, p. 15]. Scientists, of course, believe that modern scientific origin stories contain elements of truth that will survive. Astronomer Carl Sagan (1934–1996) said that the chief difference between creation myths and "our modern scientific myth of the Big Bang is that science is self-questioning, and that we can perform experiments and observations to test our ideas. But those other creation stories are worthy of our deep respect" [Sagan, 1980, p. 258].

Types of Creation Myths

Creation myths seek to reveal an absolute reality. Sproul [1979] characterizes absolute reality as a reality that is both transcendent and immanent. A transcendent reality is true for all time and space, while an immanent reality is true in the here and now. Absolute realities are proclaimed by both religions and creation myths. For example, the first book of the Old Testament proclaims an absolute reality:

> **Q9.1.** In the beginning God created the heavens and the Earth. [*Bible*, NIV, 2002, Genesis 1:1]

This proclamation has been accepted by some religions and is part of their cosmogony.

A cosmogony is a narrative that describes the establishment of the order and existence of the universe. A creation myth is a cosmogony. Closely related to cosmogony is cosmology, which is the study of the universe throughout its existence. Literally, cosmogony means creation of the cosmos and cosmology means study of the evolving cosmos. Strictly speaking, cosmogony focuses on the origin of the universe and cosmology focuses on the behavior of the already existing universe. Many theories of cosmology include a discussion of the origin of the universe and therefore incorporate cosmogony. This is especially true of modern scientific theories of cosmology that need to specify the initial conditions of the universe, as we have seen.

The essence of a creation myth is to explain how some-thing came from no-thing. Each creation myth must define what it means by "no-thing." I am drawing a distinction here between no-thing and nothing. Nothing implies the absence of physical substance. By contrast, no-thing may have physical substance but is not ordered in any meaningful way so that it cannot be characterized as a particular thing.

The literature on creation myths is voluminous. This attests to the interest human beings have in the topic of our origin. We illustrate the range of creation stories by describing a few of the categories that are used to classify creation stories [Leeming and Leeming, 1994].

Creation from Division of Primordial Unity

Mythmakers imagined that the cosmos was created by the division of a primordial unity, an analogy with the birth process. The primordial unity was the entity that existed before we did. The primordial unity could be a cosmic egg, a primal void called chaos, or original parents. When the primordial unity is represented by original parents, the sky is often the world father and the earth is the world mother. Together, the world father and the world mother are a mass of no-thing, that is, chaos; the world father and the world mother become the world when they are separated. In this context, upper-case Chaos refers to an entity in the context of a particular creation story, while lower-case chaos refers to a state of being.

Chaos appears in many creation stories. Chaos was the indeterminate, undifferentiated no-thing-ness that existed before some thing gave form and reality to Chaos. Creation in some myths means the replacement of chaos with order. Some mythmakers have included in the concept of chaos the idea that the material of creation has always existed. This idea is significant because it says that the material of the cosmos did not have to be created; it just had to be organized. By contrast, some creation stories assert that nothing existed prior to creation of the cosmos, and that the cosmos had to be created from nothing. This is often referred to as creation *ex nihilo*, or creation from nothing.

Creation by Sacrifice

Creation by sacrifice almost always involves the sacrifice of a god. Christians interpreted the crucifixion of Jesus by the Romans as the sacrifice that was needed for the forgiveness of sin. The Greek god Dionysos was the god of wine and fertility of the vineyard. Like Jesus, Dionysos was killed and then resurrected [Hamilton, 1942].

Another example of creation by sacrifice was Osiris [Bierlein, 1994], the Egyptian god of death. Osiris was born of the illicit union of the earth god Geb and sky goddess Nut, who was still the wife of the sun god Ra. Osiris was buried alive by his evil brother Set. Osiris' death by suffocation brought misery, sin, and death to Egypt. In death, Osiris ruled over the souls of people who were judged worthy by three other Egyptian gods.

Worshipers of Jesus and Dionysos believe that their resurrections assured us that death was not final. Similarly, worshipers of Osiris were given hope in the existence of life after death.

Creation from Nothing (aka creation *ex nihilo*)

Creation *ex nihilo* refers to creation from no-thing. In some *ex nihilo* creation stories, no-thing is nothing, while in other *ex nihilo* creation stories no-thing may refer to a primal void or chaos if these terms are being used to describe nonexistence. The Egyptian creation story exemplifies creation by secretion. In this type of creation story, the god figure may create by a variety of means, such as thinking, speaking, breathing, or secreting bodily fluids. There is usually a solitary male god that creates from his secretions, such as spit, vomit, semen, and sweat. The Buddhist creation story exemplifies creation by thought, where creation is a projection of the creator's thoughts.

Another example of creation from nothing is exemplified by the first creation story in Genesis[1] [*Bible*, NIV, 2002; Leeming and Leeming, 1994, p. 115]. The Book of Genesis is found in both the Jewish *Torah* and the Old Testament of the Christian *Bible*. Moses (ca. 1400 BCE) is historically considered the author and compiler of the first five books of the Old Testament. These five books are Genesis, Exodus, Leviticus, Numbers, and Deuteronomy. Together they are called the *Pentateuch* and are the five books of the *Torah*.

Among ancient religious texts, the Babylonian creation story presented in Enuma Elish predated Genesis. The Babylonian creation story outlined below may have provided the inspiration for some of the material in Genesis. Both Genesis and Enuma Elish describe a struggle between good and evil. The struggle is represented in Genesis by the relationship between God and the serpent. Even though Genesis contains some similarities to contemporary or earlier creation stories, the Genesis account of creation differs in at least one important way from the creation stories that existed at the time of Moses, ca. 1400 BCE. Most creation stories prior to Genesis were polytheistic, while Genesis was monotheistic: one true God ordered the universe and created humanity.

Genesis contains two creation stories. The creation story in the first chapter of Genesis focuses on the creation of the universe. The creation story in the second chapter of Genesis focuses on the creation of Eden, man, woman, and the knowledge of good and evil. The two creation stories in Genesis are examples of creation by word, as stated explicitly in Psalm 33:6:

> **Q9.2.** By the word of the Lord were the heavens made, their starry host by the breath of his mouth. [*Bible*, NIV, 2002, Psalms 33:6]

The first and second creation stories differ in some details. For example, the creation story in the first chapter of Genesis says that God made vegetation on the third day of creation:

> **Q9.3.** And God said, "Let the earth sprout vegetation: seed-bearing plants, fruit trees of every kind on earth that bear fruit with the seed in it." And it was so. [*Torah*, 1992, Genesis 1:11]

God made man on the sixth day of creation:

> **Q9.4.** And God created man in His image, in the image of God He created him; male and female He created them. [*Torah*, 1992, Genesis 1:26]

For comparison, the creation story in the second chapter of Genesis says that God made man from dust before any vegetation appeared on the earth [*Torah*, 1992, Genesis 2:4–7], and then said that God had already planted a garden in Eden where he put the man he made [*Ibid.*, Genesis 2:8]. Differences in the stories suggest that the two Genesis creation stories had different authors. The various translations of Genesis, or any other religious text, by different reputable sources or religious denominations make it difficult for the average reader to know which, if any, of the translations actually capture the original meaning of the text. How do we know what to believe when there are competing points of view?

Historically, religious apologists have argued in support of their beliefs. For example, Western literature contains many examples of impassioned arguments for the belief in Christianity. In his 1667 poem *Paradise Lost*, John Milton (1608–1674) described the events that led to Adam and Eve being forced out of Eden. Milton hoped to "justifie the wayes of God to men" [Milton, 1667, p. 7]. He explained Adam and Eve's behavior in large part as a consequence of the battle between God and Satan (the serpent). In the *Confessions of Saint Augustine*, Augustine of Hippo (354–430) [Augustine, ca. 400 CE] described his sinful youth and explained why he converted to Christianity. Augustine was born in a provincial Roman city in North Africa in 354 CE. His autobiographical *Confessions* are a record of the evolution of his theological beliefs. As a young man, Augustine believed in Manichaeanism and consulted astrologers. Manichaeans denied the infinite perfection of God and believed in the existence of two equal and opposite powers. As he gained more experience in life, Augustine rejected Manichaeanism and astrology. He described in *Confessions* how Plato's teachings helped him justify his belief in God.

If we recognize that many religions and religious sects exist in the world today, the evidence suggests that religious apologists have not yet presented compelling arguments for their beliefs. Apparently, no one in religion or science has yet found a set of arguments that can convince humanity to adopt a single worldview.

Point to Ponder: Is Genesis 1 an Example of Creation *ex nihilo*?

Authors have been divided on this question. For example, Armstrong [1993, p. 28] said that creation out of nothing was "an idea that was alien to the ancient world." According to Armstrong, the ancient world believed that sacred raw material had existed for all eternity.

By contrast, Leeming and Leeming [1994, p. 115] said that phrases "repeated before, during, and after acts of creation" in Genesis 1 showed "a god in total command, creating from thought and *ex nihilo.*" At the end of seven days of creation in Genesis 1, there "is nothing more to be done, to control, or to create because it is all good — creation is perfect in this story."

More recently, Walton [2009, p. 44] said that the author of Genesis 1 firmly believed "that God is fully responsible for material origins, and that, in fact, material origins do involve at some point creation out of nothing." Walton observed that the theological question of material origins was not the question he was asking. Instead, Walton was asking what sort of origins account is in Genesis 1? He concluded that "Most interpreters have generally thought that Genesis 1 contains an account of material origins because that was the only sort of origins our material culture was interested in." His interpretation of Genesis 1 led Walton [2009] to propose the Cosmic Temple Inauguration (CTI) interpretation. Walton's CTI interpretation replaced the material origins interpretation with the new interpretation that Genesis 1 is a description of functional creation rather than material creation.

Popol Vuh — The Maya Creation Story

The Popol Vuh, or the book of the people, is a creation myth of the Quiché Maya. The Quiché Maya lived in the Guatemalan highlands of Central America. They were a regional power in pre-Columbian times, that is, in the period before Christopher Columbus brought significant European influence to the Americas. According to Leeming and Leeming [1994], the original version of the Popol Vuh was destroyed by conquering Spaniards in the sixteenth century. A convert from Maya to Catholicism recomposed the Popol Vuh and translated it from the Mayan language into Latin. Spanish priest Francisco Ximenez (1666–1729) made a Spanish-language translation of the Latin Popol Vuh between 1701 and 1703.

The Popol Vuh is an example of an *ex nihilo* creation myth. This can be seen from the declaration that explained the creation of the earth:

> **Q9.5.** In the beginning there were only the creators, Tepeu and the Feathered Serpent, Gucumatz, in the void and the waters. These two sat together and thought. They glittered with Sun power. Whatever they thought and whatever they said came into being. They thought the emptiness of the void should become something and it did. "Let there be Earth," they said, and there was Earth. [Leeming and Leeming, 1994, p. 186]

The earth was created, but it was not enough. The creators wanted someone to praise them. The first animals they created could not, so they were relegated to a low position. The creators made a series of beings in human form. The first few forms were imperfect and had to be destroyed. Tepeu used a great flood to destroy most of the imperfect beings. The remaining beings with human form were chased into the woods and are known as monkeys. Eventually, the creators were able to make people.

Patricia Amlin produced an animated movie of the Popol Vuh [Amlin, 1989]. Her movie depicted the tale of the creation of the world to the victory of heroic twins over the Lords of the underworld. It is worthwhile to continue the story beyond the creation of the earth to illustrate some features of the creation myth that appear in other creation myths.

According to the Popol Vuh, the Earth was the home of the Creator mother. The underworld known as Chibalba was home to the underworld lords. The Creator Mother had two sons, One-Hunter and Seven-Hunter. One-Hunter and Seven-Hunter also had sons. They liked to play ball with their sons, but the ball playing agitated the underworld lords. They called One-Hunter and Seven-Hunter to the underworld to play ball. One-Hunter and Seven-Hunter were killed in the Chibalba by the underworld lords, and the skull of One-Hunter was hung in a dead and barren tree. One-Hunter's skull became known as the Head in the Tree. The underworld lords forbade anyone from taking the fruit of the Tree.

Little Blood, daughter of Blood Chief, learned of the Tree from her father and decided to visit the Head in the Tree. She wanted to pick the fruit of the Tree. The Head in the Tree appreciated her courage and spit in her hand, impregnating her. The Head in the Tree told her that one's essence is not lost in death; it lives on in one's sons and daughters. She was told to climb back up to Earth and to not be afraid. She did not need to fear death because in her womb lived the sons of the Head in the Tree.

Little Blood returned to Earth. In six months, her father realized that she was pregnant and was angered by her fornication. Little Blood told her father that she had not known a man; she was still a virgin. She eventually gave birth to twin sons Hunter and Jaguar Deer. The magic of her twins was used to defeat their opponents and grow corn, a very important crop for the Maya.

The Popol Vuh is a complex story. A simplified genealogy is sketched in Figure 9.1. It contains good and evil gods; a virgin birth; resurrection; sacrifice and dismemberment; and emergence from an underworld. Many of these concepts are similar to Christian concepts. Were they in the original Mayan text, or did they seep into the text when it was recomposed and translated by Christians? This is the kind of question that scholars' debate.

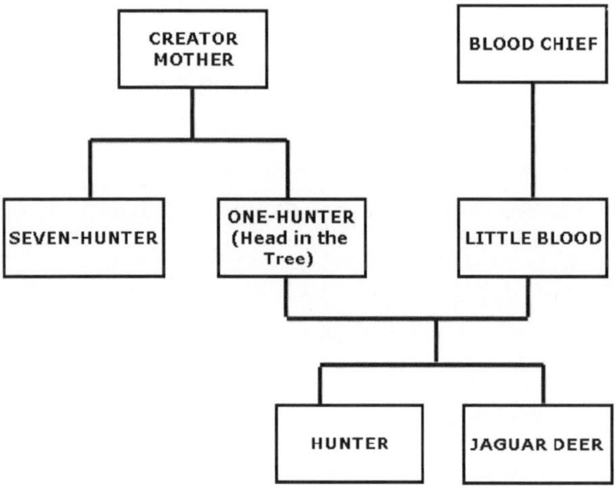

Figure 9.1. Simplified genealogy of the Popol Vuh.

Dennis Tedlock [1996] published the text of the Popol Vuh and provided a more complete discussion of the history of the Popol Vuh. The Popol Vuh is an example of pre-Colombian Central American cosmology and shows that the Maya sought to explain the origin of the universe and the origin of human beings.

Egyptian Mythology and Creation

The various centers of culture in Egyptian history modified the creation story in Egyptian mythology to give their own local deities more importance.[2] We present below a creation story from Heliopolis, an ancient cult center near Memphis that was the northern religious center of the Old Kingdom, ca. 2780–2250 BCE. Most of our information about early Egyptian creation myths is from the Pyramid Texts, a series of writings by priests from Heliopolis. The Heliopolis creation myth was the basis for later stories about the creation of the world.

According to the Pyramid Texts, Atum was the oldest of the gods. Atum was known by other names, such as the self-created, the maker of the gods, and the creator of men, but the one that is most familiar today is Ra, the Sun-God. He existed before anything else. No one knows what form he originally existed in, but he dwelt in a place called the Celestial Waters or Watery Abyss. He lived there alone until he decided to create the heavens, the celestial bodies, lesser gods, the earth, and all creatures that inhabit the earth.

Ra created his brother Shu and sister Tefnut *ex nihilo*. Ra breathed out the first air, which became Shu, god of air and the principle of life. Following Shu, Ra created the first moisture, which became Tefnut, goddess of moisture and the principle of world order. She was also the sister and wife of Shu. Ra created the Eye of Ra from the air and moisture he had made so he could see what he had created. Ra wept in happiness from his Eye and humans were created from his tears.

Shu and Tefnut, following an act of incest that was often repeated by Egyptian pharaohs, gave birth to Seb and Nut. The god Seb was the earth and his sister, the goddess Nut, was the sky. Seb and Nut were the parents of the gods Osiris, Isis, Seth, and Horus (also known as Nephthys). A simplified genealogy of the Heliopolis creation myth is shown in Figure 9.2.

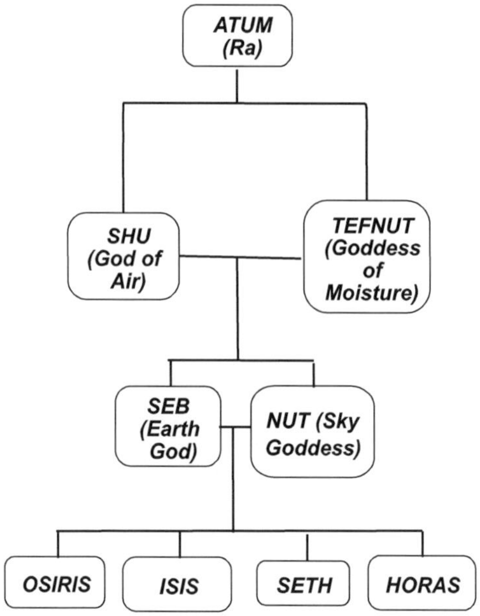

Figure 9.2. Simplified genealogy of an Egyptian creation myth from Heliopolis.

Egyptian priests from Memphis challenged the preeminence of the Heliopolitan god Atum. They asserted that their god Ptah was the original creator and Atum was subservient. Many other variations of Egyptian creation stories exist. There is a lesson to be learned from this proliferation of creation stories: if we cannot test the validity of the assertions, we will have no basis for identifying the truth. The scientific method is designed to be a process for distinguishing between competing assertions about nature. What do we use when the assertions are about the supernatural or the boundary between the natural and the supernatural?

The Babylonian Creation Story

The Babylonian epic of creation explains how the ordered universe was created from primeval chaos.[3] The epic comes to us from the Fertile Crescent, the Tigris-Euphrates Valley in what is now known as Iraq. It is a creation story that depicts good versus evil. Evil is represented by the goddess of chaos, who is symbolized by the female serpent-dragon Tiamat. Good is represented by the Babylonian god Marduk, a sun god.

The Enuma Elish is the principal source of information about the Babylonian epic of creation. Scholars are not sure when the Enuma Elish was written, but it is believed that it was written sometime during the first Babylonian dynasty (ca. 2057–1578 BCE). It appears that some of the material in the Enuma Elish dates back to the Sumerians, a culture that preceded the Babylonians in the region. The Enuma Elish is a poem that was written to praise Marduk. The poem contained an account of creation to establish a background for the story of Marduk's struggle to defeat Tiamat and achieve supremacy.

According to the Enuma Elish, the sky god Apsu and the chaos goddess Tiamat were the first gods. All of creation, beginning with lesser gods, came from the union of Apsu and Tiamat. Discord in the pantheon eventually led to confrontation between Marduk and Tiamat:

> **Q9.6.** Then joined issue Tiamat and Marduk, wisest of gods
> They strove in single combat, locked in battle.
> The lord spread out his net to enfold her
> The Evil Wind, which followed behind, he let loose in her face.
> When Tiamat opened her mouth to consume him,
> He drove in no Evil Wind that she close not her lips.
> As the fierce winds charged her belly,
> Her body was distended and her mouth was wide open.
> He released the arrow, it tore her belly.
> It cut through her insides, splitting her heart.
> Having thus subdued her, he extinguished her life.
> [Bierlein, 1994, p. 71]

Once good had vanquished evil, Marduk was able to finish the creation of the world including man. Marduk created people to serve the gods:

> **Q9.7.** I will establish a savage, 'man' shall be his name;
> Truly savage man I will create.
> He shall be charged with the service of the gods
> That they might be at ease!
> [Bierlein, 1994, p. 72]

Greek Mythology and Creation

The version of Greek mythology outlined here is from the Pelasgian creation story.[4] The Pelasgians came to Greece from Asia Minor around 3500 BCE. The first entity in ancient Greek mythology was Chaos, which corresponds to the Genesis concept of "without form." Out of the sea of Chaos arose Eurynome, the great Goddess of all things:

> **Q9.8.** In the beginning, Eurynome, the Goddess of All Things, rose naked from Chaos, but found nothing substantial for her feet to rest upon, and therefore divided the sea from the sky, dancing only upon its waves. [Sproul, 1979, p. 157]

Eurynome means "wide wandering" or "of the good name." She was the mother goddess. Her son and consort was the snake deity Ophion. Ophion is also known as the great serpent of the waters. Eurynome assumed the form of a dove and laid the universal egg. Ophion coiled seven times around the egg until it split in two. Out of the egg tumbled all of the things that exist, including the planets, the earth with its mountain ranges, and all living creatures.

Ophion and Eurynome made their home on Mt. Olympus. Eurynome created seven planetary powers and set a Titan and Titaness over each. The planetary powers were the visible heavenly objects: the Sun, the Moon, Mercury, Venus, Mars, Jupiter, and Saturn. Table 9.1 lists the relationships between the Titans, the Titanesses, and the heavenly objects.

The ancient Greeks believed that nature was unpredictable and run by capricious gods [Sagan, 1980, p. 175]. For example, Eurynome banished

Table 9.1 Greek gods, goddesses, and heavenly objects.

Titaness	Titan	Heavenly Object
Theia	Hyperion	Sun
Phoebe	Atlas	Moon
Dione	Crius	Mercury
Metis	Coeus	Venus
Themis	Eurymedon	Mars
Tethys	Oceanus	Jupiter
Rhea	Cronus	Saturn

Ophion because he took credit for creating the Universe, even though she laid the Universal Egg. The Pelasgian creation story is not the only ancient Greek creation story. It is worth considering a second creation story to illustrate how the creation stories changed over time.

The second ancient Greek creation story that we consider here is attributed to the Greek Hesiod, who was from mainland Greece during the eighth century BCE. In the beginning, according to Hesiod, there was Gaia and a great sea of Chaos. Gaia was also known as Mother Earth. Chaos produced Erebus and Night, while Gaia gave birth to Heaven and Sea. Heaven is also known as Uranus the sky.

The Titans and Titanesses were born from the union of Mother Earth and Sky. The Greek god Zeus was the son of the Titan Rhea and the Titaness Cronos. In a battle for supremacy, Zeus and his brothers Poseidon and Hades vanquished their father Cronus and their uncles, the Titans. The brothers shared power: Poseidon ruled the sea, Hades ruled the underworld, and Zeus ruled the sky from Mount Olympus. A simplified genealogy is presented in Figure 9.3.

The immortal gods decided to create mortals to inhabit the earth. Zeus ordered Prometheus and Epimetheus, sons of the Titan Iapetus, to endow the mortal creatures with gifts and powers. Epimetheus gave his gifts and weapons to members of the animal kingdom. By the time he got

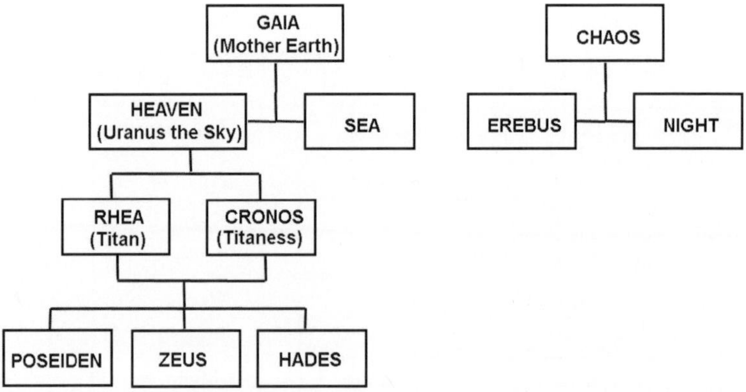

Figure 9.3. Simplified genealogy of the Greek creation myth with Gaia.

to the humans, he had nothing to give them. He left humans defenseless. Prometheus, a friend and protector of humans, stole wisdom from the goddess Athena and fire from the forge of Hephaestus. He gave Athena's wisdom to humans so they could reason and Hephaestus' fire to humans so they could be warm. When a human dies the ferryman Charon takes the human soul to Hades in the underworld.

❁ ❁ ❁

Table 9.2 is a chronology of the creation stories discussed in this chapter. Mythmakers were guided in part by what they could see and their life experiences. Nonbelievers may view the similarities between creation myths as an indication that cultures borrowed from one another over the millennia and none of the myths contain divinely inspired truth. Believers, on the other hand, may look at the same set of similarities and view the similarities as evidence that cultures in different parts of the world and at different times in history know in their souls that we all share a common heritage and face a common future.

Each creation myth presents its own concept of absolute reality. Astronomer Carl Sagan concluded:

> **Q9.9.** These [creation] myths are tributes to human audacity. The chief difference between them and our modern scientific myth of the Big Bang is that science is self-questioning, and that we can perform experiments and observations to test our

Table 9.2 Chronology of a few creation stories.

Creation Story (source)	Circa
Greek (Hesiod)	800–700 BCE
Genesis (Moses)	1400 BCE
Babylonian (Enuma Elish)	2100–1500 BCE
Egyptian (Heliopolis)	2800–2200 BCE
Greek (Pelasgian)	3500 BCE

> ideas. But those other creation stories are worthy of our deep respect. [Sagan, 1980, p. 258]

We now focus our attention on the concepts proclaimed by widely recognized modern religions. We begin with a discussion of the possible origin of religion, and then consider some widely accepted religious answers to questions about the cosmos and the existence of god.

Chapter 10

Origin of Religion

Q10.1. Science, as Justice [Oliver Wendall] Holmes was fond of saying, makes major contributions to minor needs. Religion, whether or not it comes up with anything, is at least at work on the things that matter most. [Smith, 1965, p. 11]

Traditional religion can be defined as the link between the finite and the infinite, between our mortality and an often-sought immortality. It is an attempt by humanity to strive toward something that gives meaning to our existence. Four typical functions[1] of traditional religions are summarized in Table 10.1. Traditional religious beliefs are closely held, personal relationships between an individual and one or more supernatural beings. Supernatural beings have been known by many names ranging from Allah to Zeus. Deity and god are generic terms for a supernatural being.

Relationships between people and their gods have been nurtured by a form of communication known as prayer. Prayer is an example of a religious practice. Religions usually exhibit six traits: authority, ritual, faith, tradition, mystery, and sovereignty of their principal god or gods. Religious cosmologies are based on authority and provide an explanation of the universe.

The history of the origin of religion is summarized in this chapter, followed by a discussion of the evolution of religious traits. The evolution of a religious concept is illustrated by reviewing the history of hell. We then consider three factors that are implicit in any worldview: language, logic, and faith.

Table 10.1 Typical functions of traditional religions.

Explanation	Help us understand the universe and our role in it
Exhortation	Provide instruction on how to behave (morality and ethics)
Consolation	Provide comfort in times of trial, such as illness and death
Inspiration	Give meaning and purpose to life

A Brief History of Religious Origins

The study of the origin of religion began most earnestly in medieval and post-Renaissance Europe. At the time people believed that the Genesis account of creation was correct. According to Genesis, Adam and Eve had knowledge of the divine provided directly through revelation. The purity of their beliefs was corrupted by sin and a fall from grace. Many philosophers believed that idolatry and religious diversity were produced by priests and ignorance. Darwin's theory of evolution and the growth of a critical science of history led to a reconsideration of the origin of religion.

English anthropologist Edward B. Tylor (1832–1917) introduced the concept of animism in his book *Primitive Culture* [Tylor, 1871]. Tylor suggested that the appearance of dead people in a person's dreams was interpreted by the ancients as a sign that people do survive after death but in a transformed state. This state of being was considered spiritual because of the ethereal, immaterial nature of dreams. Tylor's views became the foundation of animism. A major problem with Tylor's animism was that he could not prove a connection between the traits of primitive religions and the interpretations of dreams or visions by primitive peoples.

Animatism, an alternative to animism, was a term introduced by Robert R. Marett (1866–1943). In an 1899 talk that was later published in the journal Folklore [Marett, 1900], Marett suggested that primitive people developed the idea of a personal soul only after they had conceived an impersonal force animating the world. Marett's impersonal force came to be known as "mana," which was a Pacific Melanesian word. Later studies of native Melanesians showed that "mana" referred to a quality in spirits and people that gave them distinction; it did not refer to an inanimate force. This discrepancy cast doubt on Marett's animatism.

Another view of the origin of religion was presented during this period. James G. Frazier (1854–1941) suggested in 1890 that religion was the central step in a three-step evolutionary ladder leading from magic to science [Frazier, 1890]. He argued that people used magic to influence their environment. The failure of magic was attributed to supernatural influence. Supernatural influences are part of the province of religion. When religion was not responsive, meaning people did not get what they asked for in prayer, an alternative means of influencing the environment was sought. The lack of responsiveness of religion became, according to Frazier, the motivating force driving the development of science. Like his predecessors, Frazier's theory had a problem: there was no historical evidence that the development of magic preceded the development of religion. Frazier's evolutionary sequence could not be indisputably established.

The views of religious origins expressed by Tylor, Marett, and Frazier had a supernatural entity as the object of their worship. Old Testament scholar John Walton observed that "the distinction between 'natural' and 'supernatural' is not readily evident in the Old Testament and its world" [Walton, 2009, p. 114]. Walton said that the ancient world attributed everything to a deity.

Point to Ponder: Pie or Layer Cake?

According to John Walton, "Many people who feel caught in a perceived origins conflict between the *Bible* and science subconsciously think of the origins question as a pie. Various aspects of origins are evaluated to decide whether God did it or a naturalistic process could be identified. The 'origins pie' is then sliced up with each piece either going to 'supernatural' or 'natural' causation ..." [Walton, 2009, p. 114]. As science progresses, God's portion of the pie shrinks until, eventually, the concept of a supernatural being is unnecessary.

A more biblical view is represented by a layer cake with two layers: "In this view the realm of scientific investigation would be represented in the lower layer. This layer represents the whole realm of materialistic or naturalistic causation or processes. It is subject to scientific observation, investigation and explanation" [Walton, 2009, p. 115]. The upper layer represented the work of God.

Walton realized that the pie-layer cake analogy was not perfect because the apparently distinct divide between the two layers may not actually exist between the natural and supernatural.

A totally different view was provided by French sociologist Émile Durkheim (1858–1917) in 1912 with the publication of his book *The Elementary Forms of Religious Life* [Durkheim, 2001]. Durkheim saw the origin of religion in social terms. He hypothesized that the real object of religious worship is society. This fact, argued Durkheim, has been veiled by the extensive development of religious myths and symbolism. To illustrate his views, Durkheim suggested that the totems of Australian aborigines were built as embodiments of the ideals of the clan. Durkheim did not succeed in establishing his unorthodox view that society was the real object of religious worship. On the contrary, most people believe that the object of reverence is usually something that exists beyond society.

In his 1912 book, *The Origin of the Idea of God*, Austrian priest Wilhelm Schmidt (1868–1954) popularized the theory that "there had been a primitive monotheism before men and women had started to worship a number of gods" [Armstrong, 1993, p. 23]. People initially acknowledged the existence of only one Supreme Deity known as High God, or sometimes called Sky God because of his association with the heavens. High God "created the world and governed human affairs from afar" [*Ibid.*]. High God was eventually replaced by multiple pagan gods. According to Schmidt's theory, there was monotheism that was "one of the earliest ideas evolved by human beings to explain the mystery and tragedy of life" [*Ibid.*].

A few years later, German theologian and philosopher Rudolf Otto (1869–1937) published *The Idea of the Holy* in 1917 where he gave the term "numinous" its modern sense. A numinous experience can arouse spiritual or religious emotions in human beings that suggest the presence of divinity. Otto believed that a "sense of the 'numinous' was basic to religion" [Armstrong, 1993, p. 24].

Geoffrey Parrinder (1910–2005) observed that "errors in speculations about the origins of religion have made scholars cautious" [Parrinder, 1983, p. 14] to advance theories about the origin of religions. A widely accepted answer to the question of how religions originated is not available.

Evolution of Religious Traits

We do not know how religious thought originated. One scenario has religion arising from the fears of primitive man.[2] Religion was viewed as a means of coping with an often harsh and overwhelming environment. The evolution of a community's religious traits depended on both the type and sophistication of the community. Ancient nomads revered astronomical objects because heavenly bodies were valuable navigational tools. Early agrarian communities associated religious festivals with important seasonal activities like the planting of seeds and the harvesting of crops. As a community matured, its attitude toward religious traditions changed.

Practitioners of primitive religions tried to coerce their gods to do their will. The Celts of Stonehenge fame used human sacrifices, bonfires, and sex orgies in their worship of the sun. The Celtic Samhain was the festival when the dead could gather with the living. Over the centuries Samhain has changed from its original barbaric form to a usually more civilized observance called Halloween.

More modern religions exhibit a greater tendency to persuade than to coerce. The priest's role has grown in importance as the outward trappings of a religion, its rituals, and traditions, have become more complex. The complexity has been added in large part as a means of enhancing the persuasive powers of the religion. For example, one goal of a memorial service is to help a soul reach the "promised land" by extolling the virtues of the deceased.

Religious traits and the characteristics of deities can change as a civilization evolves. Karen Armstrong observed that "When one conception of God has ceased to have meaning or relevance, it has been quietly discarded and replaced by a new ideology" [Armstrong, 1993, p. 18]. When communities were essentially extended families, the family deity was paramount. As families gathered together to form tribes, the family deities were assimilated into tribal deities. The merging of tribes led to the merging of tribal deities. The ensuing tribal deities acquired a synthesis of traits associated with their forerunners. As an example, on a relatively large scale, the Greek Olympic pantheon led by Zeus evolved, with the conquest of Greece by Rome, into the Greco-Roman pantheon led by Jupiter. With the fall of Rome, many characteristics of the Greco-Roman pantheon became part of the Holy Trinity of Christianity.

How did religious practices develop? One possibility is that the fears of primitive people were first manifested as two kinds of spirits: hostile and neutral. Later came the belief in good or "helpful" forces and the idea of "god." People sought to hide from evil spirits, such as spirits of the dead. This led to the wearing of dark mourning clothes. They also wanted to make a hostile spirit friendly. Attempts to achieve this transformation of a spirit from unfriendly to friendly were implemented in successively more complex practices.

Rituals were one of the earliest methods of persuasion. Fetishes, such as charms or luck symbols like the swastika, were also widely used.

Rituals were usually practiced by the shaman or "medicine man." Fetishes include icons such as idols and totems. Building shelters to house fetishes was an important step on the road toward establishing churches. The people who tended the fetishes became the keepers of the churches. As their time became more absorbed by religious practices, the church keepers became full-time priests. Priests, in turn, developed complicated rituals for many reasons. The complexity of the ritual required more time and effort for the uninitiated to learn. This provided a practical means of protecting priestly authority. The need to placate and influence increasingly sophisticated spirits was another important incentive underlying the development of complex rituals.

In addition to rituals, priests used their authority to establish lists of taboos. In their original use, taboos did not carry the negative connotation they are often associated with today. Instead, they were simply prohibitions against particular behaviors that were deemed undesirable within the framework of a specific religion. Examples of taboos include the Old Testament Ten Commandments, Islamic ablution before prayers, and the Five Precepts of Buddhism. We delve a little more deeply into these taboos in subsequent chapters.

It is important to remember that the above scenario is highly speculative. No one knows all of the details of how major world religions evolved. The fundamental assumption underlying the scenario just presented is that religions evolved naturally, that is, without divine intervention. An alternative hypothesis, that an extant religion is the product of supernatural design, is difficult to accept as a starting postulate of a critical theory of history because people do not know how to investigate the validity of the hypothesis. Modern methods of studying history, as well as most other intellectual investigations, are heavily influenced by the scientific method. One of the criteria of the scientific method is the requirement that hypotheses should be testable. It should be possible to either validate or falsify a hypothesis. These topics and their implications are discussed further in later chapters.

Historical Evolution of a Religious Concept: Hell

Historian A. Bernstein published an article in 1986 that described the historical evolution of the concept of Hell [Bernstein, 1986]. Hell, said Bernstein, represents the hope of the underdog that evil will be punished in the afterlife, even if it is not in this life. Concepts of Hell have varied from one society to another. These concepts have served as imperfect reflections of the societies in which they existed.

The ancient Hebrews (ca. 1000 BCE) developed the notion of a punitive Hell. The Hebrew monarchy and priesthood promoted the idea of a single underworld for everyone. This underworld, known as Sheol, was intended to be a replacement for ancestor cults. The destruction of the Northern Kingdom of Israel in 721 BCE, followed a century and a half later by the Babylonian Exile (587–538 BCE), left the Jews helpless. Their best recourse for justice was in the afterlife. The Jewish concept of Hell acquired a punitive function.

Ancient Greece became the next stage for major adjustments to the concept of Hell. The ancient Greek philosopher Plato linked punishment or reward in the afterlife with the way a person behaved while still alive. Plato believed that only two destinies were possible in the afterlife. Seekers of truth and abstainers from fleshly lusts, such as Plato, would sit among the divine. All others would spend time in the afterlife before returning to Earth to try again.

Rome, under Augustus during the time of Christ, acquired a systematic state ideology. The philosopher Virgil applied the trend of constructing political systems to the concept of Hell. Hell became organized. Virgil associated specific punishments with various classes of offenses. Thus, the punishment for murder differed from the punishment for adultery.

Early Christian contributions to the concept of Hell began with Jesus of Nazareth. Jesus originally accepted the notion of Hell as a place of torment for "fallen angels." Early New Testament texts, by contrast, suggested that the Apostle Paul did not seem to want Hell at all. Three and a half centuries later, St. Augustine argued that Hell was both permanent and the abode of everlasting damnation.

The concept of Hell did not remain stagnant even in the intellectually stultifying climate of Europe during the Dark Ages (ca. 500–1100 CE). Barbarian Germans did not believe in Hell because they did not believe either kings or gods could confine evil. The Cathars, who roamed Northern Italy and Southern France in the twelfth and thirteenth centuries, did not believe that time would end. If time did not end, they surmised, then the biblical Last Judgment would never be held. Like Plato, the Cathars believed in reincarnation.

Hell, as a real-world manifestation of a theological concept, reached its peak of influence on daily life in post-Renaissance Europe and the North American New World. Theologians, especially Roman Catholics, used Hell as a deterrent to sin. Hell was given a physical character by preachers attempting to impress their congregations. The laity's belief in Hell reached its apogee with the publication of Dante Alighieri's (1265–1321) *The Inferno* as part of *The Divine Comedy* in the period from 1308 to his death. Dante provided a poet's interpretation of Hell and used his epic poem to publicly denounce many of the people who forced him to flee his home in Florence. He showed Hell as a place of suffering where the type of punishment depended on the sin.

Rather than accepting the concept of Hell as a distinct physical reality, German theologian Martin Luther (1483–1546) argued that Hell was personal: each individual carried his own Hell within. Hell, to Luther, was a personal estrangement from God. This concept of Hell was significantly different from that espoused by New World churchmen, who stressed hellfire and brimstone. European acceptance of Hell as a legitimate theological trait was beginning to wane in the Old World.

The eighteenth-century French philosopher Voltaire (1694–1778) decried established churches as tools of oppressive governments and denounced Hell as a tool of churches. John Stuart Mill (1806–1873), a nineteenth-century English philosopher, claimed that the idea of Hell was not consistent with the notion of a forgiving God. Austrian psychoanalyst Sigmund Freud (1856–1939) thought of Hell as a sublimation of anxiety common to all humanity. Old World doubts about the legitimacy of Hell spread across the globe until, today, the concept of Hell has lost much of its divine character in modern intellectual life. Today, some theologians tend to think of Hell as separation from God and evil as the absence of God.

Contemporary theologian and well-known Christian author Max Lucado (1955–) uses passages from the *Bible* to remind people of the biblical Hell. Lucado paints a more Dante-like picture of Hell and observes that "Hell is reserved, not for those souls who seek God yet struggle, but for those who defy God and rebel" [Lucado, 2007, p. 98]. God does not want people to spend eternity in Hell, and He told people how to avoid Hell, but God must impose justice: "How could a loving God send sinners to Hell? He doesn't. They volunteer" [*Ibid.*, p. 99]. How do people volunteer? According to Lucado, people are defiant and rebellious. They refuse to follow the Biblical path to salvation that Christ provided. If Hell does, in fact, exist, its dismissal by modern opinion makers is a major mistake.

Language, Logic, and Faith

Before continuing our discussion of alternative worldviews, we consider three factors that are implicit in any worldview: language, logic, and faith.

Language

One of the most basic constraints is the use of language.[3] The linguist Rudolf Carnap (1891–1970) succinctly defined the notion of language:

> **Q10.2.** A language, [for example] English, is a system of activities or, rather, of habits, [that is] dispositions to certain activities, serving mainly for the purpose of communication and of co-ordination of activities among the members of a group. [Carnap, 1970, p. 3]

Our use of language is necessitated by our desire to communicate. English philosopher Antony Flew (1923–2010) was a famed atheist before converting to Christianity. Flew recognized the importance of language to the human condition:

> **Q10.3.** We can have no access to concepts except through study of linguistic usage and, hence, the use of those words through which these concepts are expressed. This insight recalled to me the biblical scholars … who study some peculiar Old Testament concept by collecting and examining, with as much context as they can find, all available usages of the relevant Hebrew word. [Flew, 2007, p. 38]

Language is a collection of activities — such as sounds, motions of the body, or symbols on a page — which we have learned to associate with certain objects or concepts. Thus, the letters r-e-d are associated with a color in English, yet Germans use the different set of symbols r-o-t to designate the same color. We should also realize that the symbols of a language are only representations, or models, of the objects they symbolize. Again, referring to our color example, the letters r-e-d are merely a shorthand for describing light having specific properties. If you did not know how to read, the symbols r-e-d would be meaningless even if they were printed in red! In other words, language can be used to model the world, but the model cannot replace the world. Therefore, the best we can hope to achieve when using language to discuss nature is the construction of a model which closely approximates all of the known features of nature.

Old Testament scholar John Walton understood the relationship between language and culture: "... when we read a text written in another language and addressed to another culture, we must translate the culture as well as the language if we hope to understand the text fully" [Walton, 2009, p. 9]. Walton's careful reading and interpretation of Genesis 1 led to his functional interpretation of the biblical creation story.

Point to Ponder: What Does Creation Mean?

Creation is the act of bringing something into existence. What do we mean by existence? In philosophy, existence refers to the ontological property of being, and being refers to a thing that exists. This circular definition is not particularly satisfying.

An alternative answer to the question of the meaning of creation is to note that the modern perception of creation is that something material was brought into existence. According to American physicist Stephen M. Barr (1953–), this is a bias associated with the scientific materialism of our age. Barr argued that the basic tenet of scientific materialism is that "nothing exists except matter, and that everything in the world must therefore be the result of the strict mathematical laws of physics and blind chance" [Barr, 2003, p. 256]. He concluded that the intellectual dispute between religion and science is actually a dispute between religion and materialism.

John Walton (1952–) chose to consider the meaning of creation in terms of ontology, that is, the nature of being. Walton [2009, p. 24] recognized the materialism of our modern age and defined material ontology as "the belief that something exists by virtue of its physical properties and its ability to be experienced by the senses." He observed that material creation is one example of creation. Another example of creation is functional creation.

According to Walton [2009, p. 26], functional ontology allows something to be brought into existence by "giving it a function or a role in an ordered system, rather than giving it material properties." The modern computer illustrates both material creation and functional creation. The manufacture of a computer from basic materials is an example of material creation in an ordered system. The computer will not work, however, unless the material components are told what to do. The operation of the computer is an example of functional creation.

Logic

Besides a mode of communication, a language, we are also going to need a set of rules — a logic system — from which a consensus can be developed and predictions made. A logic system, the "rules of the game," is not necessary if we are developing a highly personal worldview that does not need to withstand close scrutiny from other people. The logic system is also unnecessary if we do not intend to subject our worldview to tests by making predictions. On the other hand, if we hope to construct a widely applicable and acceptable worldview, then some rules are needed.

Acceptance of a proposed worldview, or theory, depends on how well the theory can make predictions. Logician Patrick Suppes noted:

> **Q10.4.** It is often said that the most important critical test of any scientific theory is its usefulness and accuracy in predicting phenomena before the phenomena are observed. Any such prediction must involve application of the principles of logical inference. [Suppes, 1957, p. xv]

If our worldview only explains what has happened in the past with no implications for the future, then its usefulness will be severely restricted. However, if we can make predictions and then verify them, our confidence in the worldview will grow, as will our willingness to apply the worldview to new situations. Thus, a logic system enables us to make predictions; predictions let us test our views; and successful verification of our predictions breeds confidence. Unfortunately, application of this procedure is not enough to establish a unique, unquestionably correct worldview.

Faith

One of the problems that arises in the quest to understand nature is the lack of information. Gaps in our scientific knowledge show us the limitations of our knowledge and suggest that it is unwise to have complete confidence in the theories we propound. A striking example of what can happen when we do not recognize our own ignorance occurred in the field of physics late in the nineteenth century. According to the folklore of physics, some influential members of the physics community believed that the "laws of nature" were completely known.[4] They thought that all observable phenomena could eventually be understood by application of these "laws," and believed the only work left for succeeding generations was the verification of the "laws" to higher orders of accuracy. Succeeding generations showed their "laws" to be merely approximations of far more complex and profound physical phenomena. We now know that the "laws" of late nineteenth century physicists represent a subset of more fundamental principles.

The theory of nineteenth-century physics — classical physics — was quite successful in explaining the observations they could make with the technology then available. Classical physics was based to a large extent on the mechanistic laws of Isaac Newton. There was no reason to disbelieve classical physics, with the result that contemporary scientists developed an understandable but excessive faith in their theories. They failed to recognize the presence of faith in their work, especially their belief that nature held no more surprises for them. Their error was not the faith itself, for faith is necessary when exploring the unknown, but their lack of awareness of the role faith played in their work. We do not want to repeat their mistake, so let us recognize at the outset the importance of faith.

Just when is faith necessary? Faith must be exercised when we adopt assumptions, and when we move between hierarchical levels of nature such as those depicted in Table 6.1 in the summary of the transition from inorganic chemicals to life. In the former circumstance, acceptance of a set of assumptions constitutes a belief that this particular set of assumptions will yield a valid theory. In the latter circumstance, faith is required because we do not have a continuous connection linking the different hierarchical levels, especially those considered complex. The amount of faith needed in each circumstance depends on our knowledge. For example, if we learn a particular assumption is necessary to account for certain types of observations, then the observations give us experimental justification of the assumption and our faith does not need to be strong. Ultimately, we hope to have substantial experimental justification for all of our assumptions. Until then, faith will be active in science much as it is in religion.

Point to Ponder: The Faith of Scientists

Many scientists hesitate to acknowledge the role of "faith" in their work. They fear the word connotes an ethereal quality which would cast doubt on the objective reality of their work. Yet if we recall one definition of faith as being a belief in something which cannot be proved, then acts of faith become commonplace in the day-to-day work of scientists. A noteworthy example is the scientist's faith in an important assumption: what was learned yesterday will continue to be true tomorrow. Although the evidence may be overwhelming that certain natural "laws" seem to be true over time, there is no proof they will remain so forever. A belief in the constancy of the "laws" of nature is an act of faith. One goal of our review of different worldviews is to highlight the major assumptions — the acts of faith — underlying the worldviews we discuss. An awareness of these assumptions will help us evaluate the validity of contemporary worldviews.

Religions have varied greatly over the centuries and throughout the world. They have comforted and inspired billions of people. They have given meaning to people's lives and provided answers to ancient questions about the cosmos and the supernatural. They have also been used to justify carnage and war. We now turn to a brief review of exemplary religious answers to questions of cosmological importance.

Chapter 11

Eastern Theology

Our search for answers to fundamental questions continues with a brief description of the world's major religions in this chapter and the next: Taoism, Hinduism, Buddhism, Judaism, Christianity, and Islam. Each of these religions addresses questions of cosmological importance, and it is this aspect which receives our greatest attention. Several questions of philosophical importance naturally arise out of religious teachings. Philosophical attempts to answer fundamental questions are examined in later chapters.

During the past half-century, some scientists and philosophers have called attention to similarities between the mysticism of Eastern religions and some of the concepts of modern science, especially theoretical physics.[1] To appreciate the cosmological implications of these similarities, we need to have some familiarity with the theological context of mysticism. Therefore, a brief description of three of the most important and enduring Eastern religions is provided in this chapter.[2]

Taoism

The origins of Taoism (pronounced "dow-ism") in ancient China are somewhat obscure because of the Taoist belief that no true mystic would know he was one. As a result, the authors of the basic Taoist text, *Tao Te Ching* ("The Way and It's Power" or "The Way of Life"), remain anonymous as a matter of religious principle. For a mystic to seek recognition by claiming authorship would be a violation of part of the first poem of the *Tao Te Ching*:

> **Q11.1.** The secret waits for the insight
> Of eyes unclouded by longing;
> Those who are bound by desire
> See only the outward container.
> [Lao Tzu, 1955, p. 26]

Nevertheless, authorship of *Tao Te Ching* is attributed to Lao Tzu (ca. 369–286 BCE). No one knows, however, if Lao Tzu is the name of an ancient mystic who lived between the 5th and 3rd centuries BCE, or if Lao Tzu symbolizes a mythical character. Whether or not Lao Tzu lived or authored the *Tao Te Ching* is a question that need not concern us here; only the meaning of the poems of the *Tao Te Ching* is important.

The word "Tao" may be translated as "path" or "way." Taoism is concerned with three particular paths. The path of ultimate reality can be known only through mystical insight or revelation. It cannot be translated accurately into words. If you want to know the ultimate reality — the Way — it must be by revelation, for the Way transcends the senses and cannot be conceived. But the Way does not merely transcend life; it orders life as well. The Way is the path of nature. It is *the* course of natural events. This is the second of the three paths. Just as Tao orders events in nature, so also does Tao represent how people should live: in harmony with nature. Instead of opposing nature, we should "go with the flow." Our actions will succeed if they are in harmony with the Way.

Insights into how the meanings of Tao may have evolved can be obtained by noting that the ancient mystics were probably lonesome dwellers in remote valleys of China. Under these circumstances, they would have had ample time to meditate and reflect on the processes of nature: the changing seasons; the life cycles of fellow creatures; the beauty of the night sky; the power of nature's storms. Confronted with such processes, it is not surprising that they would feel small and inconsequential by comparison. Everywhere they went they would see change. Flowers would blossom and die. The position of stars in the heavens would change with the seasons. Storms would come and go. Even the mystics themselves would change as they aged. The importance of change in nature would be obvious to these secluded people.

Yet they would also notice that many of the changes were not haphazard. The changing seasons followed a certain order. The changing positions of the moon were repeated periodically, as were the faces of the moon. A bird would lay eggs that eventually hatched and became birds who in turn laid eggs. There seemed to be an order in nature affecting even the mystics' lives.

Perhaps a mystic had not yet learned his lesson and tried to change the order of nature. He might wish to divert streams nearer to his campsite. To his consternation the now convenient stream could become a destructive rapid after a heavy rain. Or he might try rearing and domesticating wild animals only to have them disappear during the mating season. The mystic would discover that changing the order of nature could be difficult. He would find living in harmony with nature required much less effort than acting contrary to the Way of nature. Would not the Way be the most appropriate lifestyle for the recluse?

Tao refers to a process or a path; it should not be thought of as a monotheistic entity. As an impersonal process or path, Tao should be followed rather than worshiped. The concept of being one with Tao means that we are living in harmony with a natural environment that is flexible and changing.

The key to Chinese mysticism is the phrase "wei wu wei" which can be translated roughly as "to do without doing." Ideally "wei wu wei" refers to the ultimate action and freedom passing through us when we subordinate our personal wants to the Way. A physical analogy is the flowing stream. Like Tao, the stream can carry objects with apparent ease while adapting itself to its surroundings. The stream is also incomparably strong. It can, given time, erode a mountain. Despite the inherent strength, we can easily run our fingers through the water without fear of harm. Thus, the stream, like "wei wu wei," is easily adaptable yet capable of exhibiting great strength. If we passively accede to the Way, then we will be capable of great deeds much like the stream can erode a mountain. It is up to each of us to decide.

Contained within this last comment is an important point: Taoists believe we can exercise free will. Although the Way can order our lives, we are free to accept or reject the Way. The notion of free will is discussed in more detail in later chapters.

Hinduism

While the ancient Chinese were embracing Taoism, the ancient Indians were accepting the tenets of Hinduism. The spiritual source of Hinduism resides in the ancient scriptures known as the *Vedas*. A single portion of the *Vedas*, the *Upanishads*, embodies the spiritual essence of Hinduism. The most widely read Hindu text is, however, not a *Veda* but an excerpt from one of the longest poems in the world: the *Mahabharata*. The excerpt, written anonymously between the 5th and 2nd centuries BCE, is called the *Bhagavad-Gita* (translated as the "Song of God"). It records the conversation between the god Krishna and a man named Arjuna who is seeking advice regarding a difficult decision: whether or not he should lead his household and allies into battle against his kinsmen. The looming battle becomes a symbolic backdrop for Arjuna's true quest: a search for spiritual enlightenment. What advice did Arjuna receive?

Hinduism begins by addressing the wants of man. According to Hinduism, people have essentially four wants. The first and most superficial is a want to achieve pleasure, especially physical satisfaction. The second want is somewhat more sophisticated than the pursuit of physical pleasure; it is the pursuit of worldly success. This want is very familiar to many Westerners because it is a desire to attain wealth, fame, and power. Taken together, a person's search for pleasure or worldly success is known as the Path of Desire.

As we attain pleasure or wealth, we may wonder if there is more to life. Such thoughtful reflection leads to the third want of man: duty, particularly religious duty. This want represents our search for meaning and value beyond ourselves. It is a search that may lead to the fourth and most profound want of all: liberation. Here we are seeking to transcend our mortality: to escape death. These latter two wants — duty and liberation — comprise Hinduism's Path of Renunciation.

Hindus believe we can transcend our finite selves and achieve infinite being, infinite awareness, and infinite joy. The means to attaining liberation is to uncover and release the self-infinite center known as Atman hidden within each life. Only by finding the Atman can we become one with that mystical quality within which all things have both their origin and their end: Brahman. Brahman, in the incarnation of Krishna, described himself to Arjuna:

> **Q11.2.** Brahman is that which is immutable, and independent of any cause but Itself. When we consider Brahman as lodged within the individual being, we call him the Atman. The creative

> energy of Brahman is that which causes all existence to come into being. [*Bhagavad-Gita*, 1972, p. 74]

How can we find the hidden Atman and become one with Brahman? How do we attain liberation?

Hinduism acknowledges four paths to liberation. Each path applies to a particular type of person, of which there are four. Some people like to think and seek knowledge. These people are reflective and would use Jnana yoga to attain liberation. Others may prefer to exercise their hearts rather than their minds. For them, love and devotion are the keys to liberation. They would practice Bhakti yoga. Still others find it easier to lose themselves in their work. These active people may achieve liberation through Karma yoga. Finally, a path exists for people who tend to believe only what their senses reveal to them. In this case, a person must experience the Atman before liberation is possible. Raja yoga is for them.

Upon introspection you may see yourself as one of these types. Yet, from your past, you may remember having been a different type. Perhaps in your youth you meditated often. As you aged, your work took on a much greater significance. Which yoga is for you? The answer is the yoga that fits the type of person you are at each stage of your life. In other words, as a meditative youth you should practice Jnana yoga, while now as a career-oriented adult you would find Karma yoga to be more appropriate. Hinduism recognizes a combination of yogas may be necessary for each of us. This recognition stems from the Hindu belief that there are four stages in life.

The first stage is the learning stage when we acquire the tools we need for survival. Marriage marks the end of the learning stage and the beginning of the householder stage. During this stage, we reproduce and fulfill our obligations to family and society. When our first grandchild is born, we begin the third stage: retirement. Retirement provides the time we need to reflect on what we are and who we are. During retirement we, hopefully, work out a philosophy of life and achieve liberation. Not everyone makes it through the first three stages. Those that do enter the fourth and final stage: the attainment of liberation. They have entered the state known as "sannyasin"; they have achieved unity with the timeless reality underlying life and transcending the universe.

Those who do not reach sannyasin must try again according to Krishna:

> **Q11.3.** Those without faith
> In this, my knowledge,
> Shall fail to find me;
> Back they must turn
> To the mortal pathway,
> Subject still
> To birth and to dying.
> [*Bhagavad-Gita*, 1972, p. 79]

Within this verse is the concept of reincarnation. Its purpose is to give our finite selves time to discover Atman and thereby attain liberation from our mortality. Each of us must find our innermost self or soul before we can achieve unity with Brahman. If we choose not to search, which we are free to do, then we remain in the never-ending cycle of death and rebirth. Herein is the essence of individual responsibility and free will.

Hinduism expects change. On a personal level, we are expected to pass through the stages of life until we are free from worldly illusions known as "maya" and have broken the bonds of causality called "karma." Our road to liberation may follow many paths. Indeed, the Hindus believe the major religions of the world are simply alternate paths to the same destiny. It is the common destination of the many paths that gives Hinduism its monotheistic character, despite the widespread Hindu practice of worshiping many gods such as Shiva, Vishnu, and Shakti. All of the Hindu gods and goddesses are but various aspects of one absolute Brahman:

> **Q11.4.** O Arjuna, I am the divine seed of all lives. In this world, nothing animate or inanimate exists without me. [*Bhagavad-Gita*, 1972, p. 90]

Hindus believe the cyclic nature of life applies not only to life but also to all of the cosmos. The *Mahabharata* presents a time-scale for the age of the cosmos that is measured in billions and trillions of years [Hodgkinson, 2006, p. 132]. According to the *Mahabharata*, the cosmos undergoes an endless cycle of birth, growth, decline, and destruction. The Hindu creator of the cosmos is Brahma, an entity created by Brahman. The cycle of creation and destruction is then repeated throughout Brahma's lifetime. The cycle of birth, death, and rebirth is repeated forever.

Point to Ponder: Is There a Scientific Cosmological Model That Is Compatible with Hindu Cosmology?

We saw in Chapter 3 that Albert Einstein's (1879–1955) theory of general relativity allows for the possibility of open universe models and closed universe models. Open universe models begin as an event in spacetime and can expand indefinitely following the initial event. On the other hand, spacetime expansion eventually stops in closed universe models and begins to contract. Closed universe models can have a finite number of expansion and contraction phases. If the cycle of expansion and contraction repeats periodically and for all time, the closed universe model is an oscillating universe model.

A closed universe model having only one or a finite number of expansion/contraction phases begins at some spacetime point and ends at another. Since there is a beginning to this class of closed universe models, a creation mechanism of some kind is implied. By contrast, oscillating universe models do not need to be created because they have always existed and will continue to exist throughout eternity.

The cyclic behavior of expansion and contraction suggests that oscillating universe models may be philosophically compatible with the cosmology of religions like Hinduism. For example, Parrinder (1910–2005) observed that Hinduism embraces the idea "that time is cyclic, that the universe undergoes growth, decline, destruction and re-creation endlessly" [Parrinder, 1983, p. 211]. The age and cyclic nature of the cosmos are assertions about cosmology that are now subject to scientific scrutiny.

Buddhism

Around the year 560 BCE, an infant was born in northern India. This child, a baby boy called Siddhartha Gautama of Sakyas, was most fortunate.

His father was ruler of a small region near Benares at the foot of the Himalayan Mountains. The child grew to be a handsome young man with good family connections. At the age of sixteen, he married a neighboring princess whose name was Yasodhara. Siddhartha seemed to have everything: a handsome appearance; wealth; a good wife and son; and a secure future as heir to his father's throne. But Siddhartha was not happy. Something was missing from his life. So, at the age of 29, he left his family.

Forsaking both his heritage and his lifestyle, Siddhartha studied the Raja yoga of the Hindus. Still, he was not happy. He tried joining a band of ascetics. The ascetics sought greater spiritual awareness by denying themselves bodily pleasures and avoiding worldly pursuits. After six years of meditation and fasting Siddhartha was still discontented. Then, one May night, Siddhartha paused for meditation under a fig tree near the northeastern Indian town of Gaya. That night the troubled man was transformed from Siddhartha Gautama of the Sakyas to Buddha, the Enlightened One. The rest of his life was spent traveling through northern India teaching the truth that was revealed to him that fateful evening under a fig tree.

What was this great truth? According to Buddha, there were four truths. These truths are now known as the Four Noble Truths. Foremost is the recognition that life is "dukha," or suffering. Buddhism recognizes that we can grow emotionally and spiritually as a consequence of our suffering. Throughout our lives we suffer to some extent, but especially on six occasions beginning with the trauma of birth. Each of us must endure the trials of sickness. As we grow older, we may become shackled with an inescapable weakness such as loss of sight or limb, while others may be plagued by an incurable disease. Those of us who are lucky enough to avoid these burdens undoubtedly face separation from loved ones. In the case of the death of a parent or spouse, this separation is permanent. But life is not yet over. As the years pass, we gain a heightened awareness, sometimes accompanied by real fear, of our own mortality. Even then, when we near the end of life, we are not free from suffering the afflictions of decrepitude. What is the cause of this endless suffering?

The cause of suffering in life is the essence of the Second Noble Truth. It is "tanha": desire! Worst of all is the desire to seek fulfillment within self alone. As long as our desire remains, as long as we are selfish, we are bound within the never-ending causal chain known as "karma" which drives the unbroken cycle of birth and death called "samsara." Can we free ourselves from "samsara"?

According to the Third Noble Truth, we can be freed from "samsara," the cycle of birth and death, but first, we must overcome our selfish cravings. In a sharp break from Taoism, Hinduism, and Confucianism — religions that originated in the same region — Buddhism denies the reality of the individual soul. The eternal is not to be found in the individual, but in a rejection of self in favor of a desire-free life. If we can become selfless, we can transcend "samsara," and be released from the bondage of "karma." Only then can we attain the state of total liberation known as "nirvana." By attaining "nirvana," we free ourselves from suffering. How do we achieve such a wondrous state?

Buddha taught that we could overcome desire and attain "nirvana" by following the Eightfold Path. This is the Fourth Noble Truth.

Before embarking on the Eightfold Path, we must first be willing to learn by example: to associate ourselves with one who has followed the Path. Having established this "right association," we are ready to begin our journey along the Eightfold Path. The initial task is to acquire "right knowledge," that is, a belief in the Four Noble Truths. We may then exercise "right aspiration" by dedicating ourselves to achieving "nirvana." The next two steps — "right speech" and "right behavior" — tell us how to speak only the truth and with clarity. Our speech and our behavior should be governed by the Five Precepts of Buddhism:

1. Do not kill.
2. Do not steal.
3. Do not lie.
4. Do not be unchaste.
5. Do not drink intoxicants.

The first four steps of the Eightfold Path (right knowledge, right aspiration, right speech, and right behavior) help us establish our goals, our resolve, and our behavior. The next step in our journey recognizes that we live in the community of other people and we have physical needs.

Buddha was aware of a potential conflict between physical and spiritual needs. He addressed this issue with the fifth step of the Path: "right livelihood." Whatever job we hold should be the means, not the goal, of life while we seek spiritual advancement. We must approach the Eightfold Path with the expectation of a long journey. This expectation is the sixth step — "right effort" — of the Path. It complements the view we should have of our vocation as the means to a spiritual awakening. At this point, we are prepared to tackle the last two, and most difficult, steps of the Path.

To assure ourselves that we will not stray from the Path, we must be able to analyze our own behavior and attitudes with as little bias as possible. This self-analysis constitutes the step known as "right mindfulness" and prepares us for the final step, "right absorption." In this step, we practice the techniques of Hinduism's Raja yoga to achieve the ultimate state of liberation. Once "nirvana" is attained, so also is Buddhahood. Our journey has come to an end.

The Bootstrap Model and Quantum Mysticism

Twentieth-century efforts to relate the mystical beliefs of eastern religions to the concepts of quantum physics are now known as quantum mysticism [Marin, 2009]. This work was motivated in part by the need to understand the mathematics of quantum theory and explain the proliferation of submicroscopic particles in the particle zoo. Some physicists hypothesized that the concept of "elementary particle" is an antiquated idea which should be replaced [Capra, 1975]. Rather than assuming some particles are more fundamental than others, they ascribed a comparable worth to all particles. To them, particles should be thought of as partners in a dance. What is important in their view is the dance: the relationship between the particles. Dancers may come and go, but the dance itself remains the same. It is the permanence of the dance, as contrasted with the transience of the dancers, that places the dance on a more fundamental level than the dancers. In their view, the dance partners, or particles, are transient objects. The dance, corresponding to the fundamental interactions between particles, dictates the way the dancers move and respond to one another. Thus, what is important is not the existence of particles, but the way they interact. As Fritjof Capra (1939–) says:

> **Q11.5.** In the [bootstrap] world-view, the universe is seen as a dynamic web of interrelated events. None of the properties

of any part of this web is fundamental; they all follow from the properties of the other parts, and the overall consistency of their mutual interrelations determines the structure of the entire web. [Capra, 1975, p. 276]

A final understanding of nature is obtained, argue the bootstrappers, when a self-consistent explanation of nature is acquired. No part of this explanation should be considered more fundamental than any other part. The bootstrap explanation, in a very real sense, should be cyclical. One part will be used to explain another part, which may itself be necessary for understanding the first part. The explanation of nature will essentially pick itself up by the bootstraps. Thus, says Capra,

> **Q11.6.** … the complete "bootstrap" view of nature, in which all phenomena in the universe are uniquely determined by mutual self-consistency, comes very close to the Eastern world-view. [Capra, 1975, p. 279]

The bootstrap theory had some success in providing a metaphysical interpretation of scattering phenomena associated with hadrons. A view of the bootstrap theory by physicists Á. de Rújula (1944–), H. Georgi (1947–), and S. L. Glashow (1932–) — leading advocates of the quark model — summarized the status of the bootstrap model:

> **Q11.7.** Once upon a time, there was a controversy in particle physics. There were some physicists [boot-strappers] who denied the existence of structures more elementary than hadrons, and searched for a self-consistent interpretation wherein all hadron states … were equally elementary. Others [quark proponents], appalled by the teeming democracy of hadrons, insisted on the existence of a small number of fundamental constituents and a simple underlying force law. [de Rújula *et al.*, 1975, p. 147]

Within this quote is an allusion to a "simple underlying force law." Formulation of this "simple" law has eluded the combined efforts of an army of twentieth-century physicists, including Albert Einstein. What is this "simple underlying force law" which has captivated some of the greatest minds of all time?

We saw in Chapter 4 that physicists recognize four fundamental interactions: strong, electromagnetic, weak, and gravitational. Many physicists believe a theory can be formulated which will unify these four interactions into a single interaction in the Theory of Everything. The resulting lone interaction is the "simple underlying force law" mentioned in the above quote by de Rújula, Georgi, and Glashow. So far, no one has succeeded in unifying all four interactions. Furthermore, the view of the quark model as a particle theory has been superseded by quantum field theory.

❁ ❁ ❁

Threading through the above discussion of three major Eastern religions is the idea of change, of a world which is not static, but dynamic. Taoism tells us to passively accept natural change; Hinduism would have us change the cycle of birth and death that binds us by counseling us to seek liberation; meanwhile, Buddhism directs us to the Eightfold Path as our guide to liberation. The processes of change are as fundamental to Eastern religions as they are to some scientific theories. Does this point reveal a truth about nature, or is it nothing more than an aspect of the Hindu "maya," a worldly illusion?

Chapter 12

Western Theology

Change is a feature in both Eastern and Western religions, but a more interesting feature of Western religions from our perspective is the beginning. Each of the religions considered here — Judaism, Islam, and Christianity — accepts the Genesis version of Creation. This observation has important cosmological and evolutionary implications. Our goal in this chapter, as in the last, is to develop a broad theological perspective.[1]

Judaism

The *Bible*, the most widely distributed book in the world, is divided into two parts: Old Testament and New Testament. Each of the Testaments is composed of a number of parts. Taken together, the subdivisions of the two Testaments are a library of religious information. The Old Testament may be thought of as the library of Judaism, whereas the New Testament documents the teachings of Jesus of Nazareth (ca. 2 BCE–33 CE) and his followers. The first five books of the Old Testament — Genesis, Exodus, Leviticus, Numbers, and Deuteronomy — are collected in the *Torah*, a sacred text of Judaism.

The traditional materialistic view of the first line of the first book of the Jewish library — Genesis — expresses the Jewish belief that the material world was made by God:

> **Q12.1.** In the beginning God created heaven and earth. [*Torah*, 1992, Genesis 1:1; see also RSV *Bible*, 1971, Genesis 1:1]

According to the materialistic view, Judaism not only teaches that God created the material world but it also tells us what God thought of His creation:

> **Q12.2.** And God saw all that He had made, and found it very good. [*Torah*, 1992, Genesis 1:31; see also RSV *Bible*, 1971, Genesis 1:31]

By accepting the notions that the existing material world was created by God and is considered good, we implicitly attribute worth to existence. Unlike Eastern religions, which put a relatively low value on material objects, Judaism teaches that matter, including humanity, has worth. This is a distinguishing feature of Judaism, but not the most fundamental. What separated Judaism from earlier religions is the Jewish acceptance of a meaningful world.

Judaism teaches that the events of history have a purpose decreed by God and that historical events involving humanity are meaningful. The Genesis version of creation exemplifies this point. It asserts that the arrival of man[2] in the universe was planned. Attributing meaning to history gives significance to the activities, the institutions, and the social order of humanity because they can influence the course of history. Anything capable of affecting history can aid or deter the fulfillment of God's purpose.

At this point, we may argue that if God is omnipotent as Judaism claims, then how can the activities of man, being a mortal and frail creature, interfere with God's goals? Judaism recognizes the failings of man — the fall of Adam and Eve from God's grace is a symbolic example — but it also says that man is free. Each of us can choose to do God's bidding, or we can refuse. The choice, by God's grace, is ours:

> **Q12.3.** I call heaven and earth to witness against you this day: I have put before you life and death, blessing and curse. Choose life–if you and your offspring would live–by loving the Lord your God, heeding His commands, and holding fast to Him. [*Torah*, 1992, Deuteronomy 30:19–20; see also RSV *Bible*, 1971, Deuteronomy 30:19–20]

We have here an illustration of a potential confrontation between free will and determinism. Evidence of this clash was noted in our discussion of Eastern religions. The clash takes on a more poignant meaning in the context of Western theology. If an omnipotent being grants us freedom, could not that freedom be just as easily removed? Are we truly free, or is our freedom limited in some way? If it is limited, what are those limits and when do they deter our actions? These questions arise not only in Judaism, but in any religion that believes in an all-powerful god yet says man's will is free. Possible answers to these questions and their implications are discussed in subsequent chapters.

Suppose we accept the Jewish notion that there is meaning in history. What kind of social order should we establish if we want to fulfill God's purpose? Although several social orders had previously existed, such as the comprehensive social structure developed by Confucius in ancient China, the Jewish social order and the relationship between God and His people were documented in the book of Leviticus. A set of laws was brought down from a mountain called Sinai by a man named Moses (ca. 1400 BCE). The set of laws is known as the Ten Commandments:

I. No gods shall be exalted more than God;
II. Idols should not be worshiped;
III. Do not use God's name in vain;
IV. Remember the Sabbath;
V. Honor your parents;
VI. Do not kill;
VII. Do not commit adultery;
VIII. Do not steal;
IX. Do not bear false witness against your neighbor; and
X. Do not covet your neighbor's wife or possessions.

The Ten Commandments address the four major causes of an unstable social order: power, money, sex, and language. Unlimited power or wealth can easily, though not necessarily, corrupt, as can the human desire to attain power and wealth, a desire known as greed. Sodom and Gomorrah are Old Testament examples of social systems in which greed and sexual permissiveness flourished. But the causes of corruption are not just power, money, and sex. Language, if completely unbridled, can create great suffering. This has long been recognized by society. Even in the United States, where freedom of speech and the press are Constitutionally guaranteed, it is illegal to whimsically yell "fire!" in a crowded theater. These constraints are needed if people wish to live together peacefully. The last six of the Ten Commandments provide a foundation upon which a stable social order can grow and prosper. Their function is comparable to the Five Precepts of Buddhism.

Beyond establishing a minimum framework for constraining the relationship between one person and another, the first four of the Ten Commandments decree what God expects of us in relation to Him. Foremost of these Commandments is the following decree:

> **Q12.4.** You shall have no other gods besides me. [*Torah*, 1992, Exodus 20:3; see also RSV *Bible*, 1971, Exodus 20:3]

God will accept no rivals. In return God vowed to show

> **Q12.5.** … kindness to the thousandth generation of those who love Me and keep My commandments. [*Torah*, 1992, Exodus 20:6; see also RSV *Bible*, 1971, Exodus 20:6]

Thus, the pact between man and his Creator was set.

Point to Ponder: What Is the Cosmic Temple Inauguration Interpretation?

We have indicated in previous chapters that the materialism interpretation of Genesis 1 presented in this section is not the only possible interpretation. Old Testament scholar John Walton (1952–) has argued that the materialism interpretation of Genesis 1 should be replaced by a functional interpretation [Walton, 2009] that he called the Cosmic Temple Inauguration (CTI) interpretation. According to Walton, this label "picks up the most important aspect of the view: that the cosmos is being given its functions as God's temple, where he has taken up his residence and from where he runs the cosmos" [*Ibid.*, p. 162]. He said that the most distinguishing feature of the CTI view was "the suggestion that, as in the rest of the ancient world, the Israelites were much more attuned to the functions of the cosmos than to the material of the cosmos."

Walton identified several key features [Walton, 2009, p. 162] of the CTI view:

- The Hebrew word translated "create" (bārā') concerns assigning functions.
- The account begins in verse 2 with no functions (rather than with no material).
- Days 1–3 pertain to the 3 major functions of life: time, weather, and food.
- Days 4–6 pertain to functionaries in the cosmos being assigned their roles and spheres.
- The recurring comment that "it is good" refers to functionality (relative to people).
- The temple aspect is evident in the climax of day 7 when God rests — an activity in a temple.

If the CTI interpretation is correct, the view that the timeline in Genesis 1 refers to the material creation of the cosmos is no longer valid because the CTI interpretation says that the timeline refers to assigning functions, including the creation of time.

Walton concluded that the account in Genesis 1 "can be seen to be a 7-day inauguration of the cosmic temple, setting up its functions for the benefit of humanity, with God dwelling in relationship with his creatures" [Walton, 2009, p. 163].

Christianity

The Old Testament established a pact between man and God, yet man did not live up to the pact. As the Old Testament story of Adam and Eve symbolizes, man was a sinner. Could man reconcile himself to God? If so, how? Christianity accepts that people are sinners and teaches that we can attain reconciliation — or Atonement — but not by our own deeds. We need help, and that help was sent by God in the form of the Messiah.

Who was the Messiah? When we stop to ponder this question, we realize very little concrete information exists about Him (here I adopt the Christian practice of capitalizing the first letter of pronouns referring to the Messiah).

We do not have a physical description of the Messiah, nor do we know many of the details of the Messiah's life, especially His youth. What we do know of the Messiah's words and deeds is contained in the New Testament of the *Bible*, particularly in the *Gospels*, the books of Matthew, Mark, Luke, and John. Other information about the Messiah and the time in which He lived can be gleaned from history, for Christianity is a historical religion. This means the credibility of Christianity is linked to the historical accuracy of documented Christian events. The most important of these events was the life of a Jewish carpenter, born of a virgin during the reign of Herod the Great in a small Palestinian town called Bethlehem.

This man, Jesus of Nazareth, lived a life on Earth that spanned less than 35 years. He traveled often but not far. During his travels — a one to three-year period at the end of his life — Jesus performed deeds Christians believe were miracles. It was not his deeds, however, but his teachings that are of greatest importance to Christianity.

Jesus taught about love. Jesus spoke of God's love for man, and He spoke of the importance of our acceptance of God's love. Beyond this, Jesus taught that we should do more than just accept God's love; we should let it pass through us to our neighbors. Those of us who have ever had inconsiderate neighbors know how difficult it can be to love our neighbors. This is the challenge of Christian love; to care for your neighbor according to their needs, not according to their due. God's love for us is like this, said Jesus, so should we not try to emulate Him? It is not an easy task but, according to Christianity, it is the right task. It is a task Jesus fulfilled despite the harsh treatment He endured at the hands of His "neighbors," the occupants of the Holy Land in the Middle East.

As Jesus traveled and taught, He gathered round him followers. Among the followers was a small band of men known as the Disciples. It was the Disciples, not Jesus, who first asserted that Jesus was the long-awaited Messiah. Such assertions naturally aroused the suspicions of the Jewish establishment. With the passage of time, Jesus' reputation and stature grew. Some Jewish leaders — the Pharisees — began to perceive Jesus as a threat. Jesus was accused of blasphemy, but this crime was not punishable by death according to Roman law, the prevailing law. To permanently silence Jesus a new charge was needed. The charge was treason, and the sentence was death. Jesus was crucified on the cross. His suffering was immense at the hands of those He sought to help. Was He bitter and full of hatred at the very end? Far from it. Instead, He asked forgiveness for His persecutors as His mortality drew to a close:

> **Q12.6.** Then Jesus said, "Father, forgive them; for they do not know what they are doing." [The *Gospels*, 1996, Luke 23:34; see also RSV *Bible*, 1971, Luke 23:34]

As he suffered, so Christianity teaches, Jesus shouldered the full weight of human sin. At the point of deepest agony and greatest separation from God, He exclaimed

> **Q12.7.** My God, my God, why have you forsaken me? [The *Gospels*, 1996, Matthew 27:46; see also RSV *Bible*, 1971, Matthew 27:46, Mark 15:34, Psalm 22:1]

Yet the agony did not last. Jesus finally rejoined His Father:

> **Q12.8.** Father, into your hands I commend my spirit! [The *Gospels*, 1996, Luke 23:46; see also RSV *Bible*, 1971, Luke 23:46]

The Messiah, Jesus of Nazareth, died during the reign of Pontius Pilate.

Ordinarily, the story of a man's life ends at his death. In Jesus' case, His death was but a milestone beyond which His real power blossomed. For within three days Jesus Christ rose from His grave and completed our atonement with God. Never again would people need to fear death. Christ lived!

Are you finding yourself wondering, "If only this story was true," or "How much of this can I believe?" If you are skeptical, you are not alone. Jesus is considered to be a prophet by such groups as the Jews, Hindus, and Muslims. Is there any way to know whether or not Jesus is the Messiah, the personification of God? Did He perform His miracles? Was He in fact resurrected from the dead?

The *Gospels* are the source of much of our knowledge of the life of Jesus. There are other sources, however, that were not disciples of Jesus. For example, Roman historian Tacitus lived from ca. 56 to 120 CE. He talked about Christians in his Annals when he described the great fire in Rome during the reign of Nero. According to Tacitus:

> **Q12.9.** … all human efforts, all the lavish gifts of the Emperor, and the propitiations of the gods, did not banish the sinister belief that the conflagration [Roman fire] was the result of an order. Consequently, to get rid of the report, Nero fastened the guilt and inflicted the most exquisite tortures on a class hated for their abominations, called Christians by the populace. Christus, from whom the name had its origin, suffered the extreme penalty during the reign of Tiberius at the hands of one of our procurators, Pontius Pilate … [Tacitus, 109 CE, Book Fifteen of the Reign of Nero, p. 344]

This passage was written by a noted Roman historian, a believer in the Roman pantheon of gods, who lived shortly after the time of Jesus. Tacitus was not a Christian; he was a historian who confirmed that a person called Christus lived, died under the rule of Pontius Pilate, and spawned a group of followers known as Christians. Another historian, First-century Jewish historian Flavius Josephus (ca. 37–100 CE), also documented the life of Jesus [Josephus, ca. 93 CE, Chapter III]. Scholars continue to study the veracity of the Old and New Testaments and may be able to tell us more as their work progresses.[3] For now, acceptance of Jesus as Christ based on what we now know must be an act of faith. And faith, as we have already seen, appears to be an inescapable part of our quest to answer the fundamental question: does god exist?

Islam

A conflict dating from the time of Abraham casts light on a part of the modern world which continues to experience unrest: the Middle East. According to Genesis, God used water to destroy all life on Earth except for a small coterie assembled on an ark built by a man named Noah. After 40 days and 40 nights of rain, the floods subsided and it was left to Noah and his family to repopulate the Earth. One of the earlier descendants of Noah's son Shem was Abraham.

Abraham played a special role in God's plan. His role was not a simple one. God promised to make Abraham and his descendants a great nation if Abraham would submit to God's will. Abraham agreed. A slight problem loomed on the horizon, however. Sarah, Abraham's wife, was unable to conceive.

Following custom, Sarah sent her maid, Hagar, to Abraham. Hagar was able to bear children, and she looked with contempt upon Sarah. It should come as no surprise that Sarah was enraged. With Abraham's permission, Sarah dealt harshly with Hagar. So harshly, in fact, that Hagar fled.

An angel of the Lord eventually found Hagar and righted the wrong Sarah had inflicted upon her by giving Hagar a son Ishmael. But God also had an agreement with Abraham. True to His word, God gave Abraham a son by Sarah whom they named Isaac. The descendants of Isaac are the Jews, and the descendants of Ishmael are the Muslims: the followers of Islam. This is the point where the *Bible* and the Holy Book of Islam — the *Koran*[4] — begin to differ.

Islamic theology dates from the sixth century CE and is based on the *Koran*, the Holy Book of Islam written by the prophet Muhammad (ca. 570–632 CE). Muhammad was born into a Bedouin society in which there thrived a polytheistic religion with virtually no moral constraints. Both of Muhammad's parents were dead by the time Muhammad was six years old. The youth went to live with his uncle, who was not rich. Though the lad had to work as a shepherd, he experienced the glow of acceptance emanating from his new home. As Muhammad matured, the widespread immorality filled him with revulsion. He sought a better way.

Muhammad met and eventually married Khadija, a wealthy widow who was fifteen years his senior. Their relationship was based on mutual admiration and love. Khadija provided a secure home from which the troubled young man could intensify his spiritual search.

Muhammad's favorite spot for meditation was a cave in a mountain called Mount Hira, near Mecca. This cave was where he first realized that there was but one god, and that god is Allah:

> **Q12.10.** ***La ilaha illa Allah!*** (There is no God but Allah!)
> [Muhammad in Smith, 1965, p. 221; and Smith, 1991, p. 225]

Muhammad was commissioned a prophet by a voice from heaven in his mountain cave. The voice demanded he cry out, but what should he cry? The voice responded:

> **Q12.11.** Recite in the name of your Lord who created — created man from clots of blood. Recite! Your Lord is the Most Bountiful One, who by the pen taught man what he did not know. [Muhammad in the *Koran*, 1996, 96:1]

The revelations from Allah were delivered to Muhammad by the voice of the angel Jibril, (also known as Gabriel). The only miracle Muhammad ever claimed was his recitation of the revelations for compilation in the *Koran*.

Muhammad's message was egalitarian: all of us are equal in the sight of the Lord. This message was not well received by the society in which he lived. Muhammad was forced to flee Mecca. His flight, known as the Flight of the Hegira, took him to Medina where he became a leading statesman. In time, he returned to Mecca as its conqueror. The use of force to punish wrongdoers is consistent with the teachings of Islam in the *Koran*:

> **Q12.12.** Permission to take up arms is hereby given to those who are attacked, because they have been wronged. God has power to grant them victory: those who have been unjustly driven from their homes, only because they said: 'Our Lord is God.' Had God not defended some men by the might of others, the monasteries and churches, the synagogues and mosques in which His praise is daily celebrated, would have been utterly destroyed. [Muhammad in the *Koran*, 1996, 22:39–40]

Upon his triumphant return to Mecca, Muhammad did not seek vengeance. Instead he treated the conquered, his former persecutors, with compassion. The entire Meccan population was converted to Islam. Two years later Muhammad returned to Medina and died. What had this man, whose behavior was uncharacteristically civilized and whose achievements were remarkable, learned during his lonely vigil those many nights in a mountainside cave? What was the inspiration that led him to prevail over his barbaric world?

The answer is contained in the *Koran*, a book not quite as long as the New Testament. The *Koran* was recited to Muhammad by the voice of the angel Gabriel and is the basis of Islam.

Allah is the god of Islam. He is an all-powerful, all-pervasive, and benevolent god, just as the Jewish god Jehovah is. In fact, Muslims believe Allah and Jehovah are synonymous. The followers of Islam contend that the Old Testament and the New Testament are incomplete. They argue that the two Testaments are older than the *Koran* and were therefore written at an earlier stage of man's spiritual development.

Even though there are differences between the Testaments and the *Koran*, there are also similarities. Islam, like Judaism and Christianity, is monotheistic. Islamic acceptance of Genesis implies acceptance of the view that existence has worth. Furthermore, Islam teaches that man is an individual capable of exercising free will. Indeed, when the Day of Judgment arrives, Islam says each of us, each of our souls, will have to account for ourselves. How well we have adhered to the tenets of Islam will determine how happy or miserable we will be in the afterlife. Consequently, it is important to know what the tenets of Islam are.

According to Islam, there are five principal tenets that must be observed. They are known as the Five Pillars of Islam. First of the five is our acceptance of the creed:

> **Q12.13.** There is no god but God, and Muhammad is His Prophet. [Smith, 1965, p. 236; and Smith, 1991, p. 244]

A Muslim must recite this creed at least once in his life with full understanding of what he is saying.

The second Pillar strikes at a problem we have seen before: man's inflated perception of himself. Muslims should face Mecca five times a day and pray. This is a constant reminder to them that they are not God. According to the *Koran*, this is the most difficult lesson man has to learn.

Islam also teaches the virtues of charity. Muhammad recommends two and a half percent (2.5%) of everything we own should be given by those who have material resources to those who do not. Thus, Islam tries to ease the burden of impoverished souls through transfer payments. Charity is the third Pillar of Islam.

The fourth Pillar is the Islamic insistence that Muslims fast during the month when Muhammad received his initial commission as a Prophet. This period of fasting, known as "Ramadan," is observed for a number of reasons. It teaches self-discipline; it helps those who need not suffer understand and empathize with those who must; and it reminds us of our dependence on God's bounty.

The fifth and final Pillar is Islam's demand that every able Muslim journey at least once in a lifetime to Mecca. Again, the purpose of this Pillar is to serve as a reminder. In this case, man is reminded of his equality with other men since every pilgrim, the powerful and the weak, must shed all outer trappings of social distinction. The garments worn by each pilgrim are the same as those of every other pilgrim when they enter the Sacred House at Mecca. More important than this egalitarian purpose is the heightening of the pilgrim's devotion to God and the teachings of His prophet Muhammad.

Jesus in Islam

Christianity and Islam have significantly different views of the life of Jesus. According to New Testament Professor Neil Lightfoot [2003], the central events in the life of Jesus were recorded in the *Gospels* written by Matthew, Mark, Luke, and John during the period 50–100 CE. The *Gospels* were written to help explain the significance of the birth, life, ministry, crucifixion, and resurrection of Jesus written by people who lived in the same century as Jesus. The *Koran* contains revelations about the life of Jesus delivered to Muhammad by an angel. Moslems view the *Koran*, which was written in the sixth century CE, as a more current and reputable source. Nessim Joseph Dawood (1927–2014), translator of the *Koran*, observed that

> **Q12.14.** The *Koran* accuses the Jews of corrupting the Scriptures and the Christians of worshipping Jesus as the son of God, although He had expressly commanded them to worship none but Him. Having thus gone astray, they must be brought back to the right path, to the true religion preached by Abraham. This was Islam–absolute submission or resignation to the will of God. [Nawood in the *Koran*, 1996, p. 2]

In the above quote, the word "God" refers to the Islamic god Allah. Later we see the term "Messiah" referring to a mortal rather than a god. Readers of sacred texts and their translations need to be aware that the meaning of a word or phrase may vary from one religion to another.

We describe the Islamic view of Jesus by beginning with the birth of Jesus. Islam views Jesus as the son of Mary:

> **Q12.15.** ... And you shall recount in the Book (*Koran*) the story of Mary: how she left her people and betook herself to a solitary place to the east.
>
> We sent to her Our spirit in the semblance of a full-grown man, And when she saw him she said: 'May the Merciful defend me from you! If you fear the Lord, leave me and go your way.'
>
> 'I am your Lord's emissary,' he replied, 'and have come to give you a holy son.'
>
> 'How shall I bear a child,' she answered, 'when I have neither been touched by any man nor ever been unchaste?'
>
> 'Thus did your Lord speak,' he replied. 'That is easy enough for Me. He shall be a sign to mankind and a blessing from Ourself. Our decree shall come to pass.'
>
> Thereupon she conceived him, and retired to a far-off place. [*Koran*, 1991, 19:12–22, p. 215]

Mary gave birth and took the baby to her people:

> **Q12.16.** Carrying the child, she came to her people, who said to her: 'Mary, this is indeed a strange thing! Virtuous woman, your father was never a whore-monger, nor was your mother a harlot.'
>
> She made a sign to them, pointing to the child. But they replied: 'How can we speak with a babe in the cradle?'
>
> Whereupon he spoke and said: 'I am the servant of God. He has given me the Book and ordained me a prophet. His blessing is upon me wherever I go, and He has exhorted me to be steadfast in prayer and to give alms as long as I shall live. He has exhorted me to honor my mother and has purged me of vanity and wickedness. Blessed was I on the day I was born,

> and blessed I shall be on the day of my death and on the day I shall be raised to life.'
>
> Such was Jesus, the son of Mary.
>
> [*Koran*, 1991, 19:28 ff., pp. 215–216]

Jesus had the power to perform miracles, but he was not divine; he was mortal like other prophets:

> **Q12.17.** Unbelievers are those who declare: 'God is the Messiah, the son of Mary.' Say: 'Who could prevent God, if He so willed, from destroying the Messiah, the son of Mary, his mother, and all the people of the earth? God has sovereignty over the heavens and the earth and all that lies between them. He creates what He will and God has power over all things.' [*Koran*, 1991, 5:17, p. 81]

According to the *Koran*, Jesus was not crucified, died, buried, and resurrected. Instead, Allah raised Jesus up to Heaven while unbelievers

> **Q12.18.** denied the truth and uttered a monstrous falsehood against Mary. They declared: 'We have put to death the Messiah, Jesus the son of Mary, the apostle of God.' They did not kill him, nor did they crucify him, but they thought they did. (He was made to resemble another for them.)
>
> Those that disagreed about him were in doubt concerning him; they knew nothing about him that was not sheer conjecture; they did not slay him for certain. God lifted him up to Him; God is mighty and wise.
>
> [*Koran*, 1991, 4:157 ff., p. 76]

❁ ❁ ❁

Clearly, our review in the last two chapters has barely scratched the surface in providing an understanding of humanity's greatest religions of East and West. We have seen the emergence of a monotheistic theme. Judaism, Christianity, and Islam acknowledge the existence of one god, but that one god has different characteristics. Sacred texts such as the *Torah*, the *Gospels*, and the *Koran* have been written to reveal the nature of god to humanity. If god exists, what is the nature of god?

Our comparison of Western religions has shown an acceptance of the notion that a single all-knowing (omniscient), all-powerful (omnipotent) deity coexists with human beings endowed with free will. How can these seemingly incompatible beings coexist? Are we truly free if our freedom must be granted by another being?

We have also seen that religions differ in their descriptions of the origin of the universe. Some Eastern religions speak of a cyclic universe where people live and die many times. By contrast, the Western religions we discussed assert that humans are born once and no more. Are any of these views correct?

Humanity has entered an era where possible answers to these and other fundamental questions can be expressed in a form suitable for scrutiny under the critical glare of scientific inquiry. The scientific worldview has a major influence on public opinion and on privately held beliefs. As we pointed out in Chapter 1, surveys have found that mainstream scientists like the new atheists are nonbelievers. Are nonbelievers right or have they made a mistake that will haunt them and their followers for eternity?

Chapter 13

Beyond the Natural

Our search for an answer to the question of god's existence has led us to review several worldviews, including the scientific worldview in Part 1, mythological creation stories in Chapter 9, and religious creation stories in Chapters 11 and 12. The next step in our quest is to consider key questions about worldviews from a metaphysical perspective.

Ancient Greek Metaphysics

The branch of philosophy known as metaphysics is concerned with answering questions that extend beyond the limits of the observable universe. Questions of metaphysical interest are questions which cannot be answered by science. A precise definition would characterize metaphysics as the branch of philosophy consisting of ontology and cosmology.

Questions of metaphysical interest include "Do I exist?" and "Is there a God?" These questions are similar to those discussed during our introduction to the religions of the world. The overlap of theological and metaphysical topics is part of the branch of metaphysics called ontology. Ontology is concerned with discovering the essence of being or the essence of reality.

Other metaphysical questions are concerned with ascertaining how the universe evolved and what the constituents of the universe are. Questions like these represent an overlap of science and metaphysics that is part of the second branch of metaphysics: cosmology. Cosmology encompasses the efforts of people who try to discern the structure and laws of nature.

An improved appreciation of the complexities and subtleties of modern scientific theories and their impact on contemporary worldviews is obtained by examining classical philosophical concepts. We begin in the birthplace of many metaphysical concepts familiar to the Western mind: ancient Greece.[1]

Heraclitus

One of the most important goals of early Greek metaphysicists was the characterization of the permanent features of nature. Many people throughout history have yearned for something permanent and unchanging. It is somewhat surprising, then, that an early Greek philosopher of substantial repute — Heraclitus of Ephesus (ca. 535–475 BCE) — was a proponent of the notion that nature was always changing.

Heraclitus taught that change is the only reality. Change was the only permanent characteristic observable to the ancients. The constant state of flux was interpreted by Heraclitus to mean that permanence in nature was unknowable.

The concept that change is the only reality was not unique to Heraclitus in the ancient world, nor did it originate with him. The notion that "all things are flowing" was present in the teachings of the Taoist Chuang Tzu (ca. 369–286 BCE): "In the transformation and growth of all things, every bud and feature has its proper form. In this we have their gradual maturing and decay, the constant flow of transformation and change" [Capra, 1976, p. 102]. In addition to Chuang Tzu and Heraclitus, another prominent proponent of impermanence as a dominant feature of nature was Buddha. Huston C. Smith (1919–2016) stated that Buddha was so struck "with impermanence that he listed it as the first of his Three Signs of Being, or characteristics that apply to everything in the natural order, the other two being suffering and absence of a permanent soul" [Smith, 1965, p. 129]. Change in nature and the influence of change on philosophical thought were not limited to a particular geographic locale.

Parmenides and Zeno

The influence of Heraclitus' views did not remain unchallenged. Parmenides (ca. 515 BCE) and his student Zeno (495–430 BCE) were from Elea, a Greek colony in Italy. The Eleatics, notably Parmenides, believed that things do exist, whereas change is illusory. According to Parmenides, reality or being must be permanent and unchanging: "One path [to the truth] only is left for us to speak of, namely that IT IS. In this path are very many tokens that what is, is uncreated and indestructible; for it is complete, immovable, and without end" [Stroll and Popkins, 1972, p. 206]. Only that which is unchanging can be real. What exists, or Being, cannot come from something else because there is nothing else. If this conclusion is valid, how do we account for the perpetually changing world we see?

Parmenides answered this question by appealing to mystical views like those in Eastern philosophy. Since Being is permanent and unchanging, the perpetually changing world must be unreal or illusory. To find reality, Parmenides argued, we must disregard the illusory world of appearance. This view closely resembles the Hindu view: "A world that is MAYA, [is] deceptively tricky in that it passes off its multiplicity and welter of dualities as ultimate whereas these are in fact provisional only" [Smith, 1965, p. 85]. The highest achievement of Hinduism is to escape maya and become one with Brahman. But what is Brahman? According to Huston Smith, "The name the Hindus give to the supreme reality is BRAHMAN, from the root BRIH meaning 'to be great'" [*Ibid.*, p. 72]. The Hindu and Eleatic views of nature have conceptual similarities.

The Eleatic school of thought believed that things could exist. Perhaps the greatest problem with the Eleatic view was their complete rejection of the existence of change in the world. Some middle ground between the Eleatic view of permanent, unchanging reality and Heraclitus' view of perpetual change needed to be found. The atomists provided this middle ground.

Atomists

The Eleatics believed in the existence of an indestructible substance and rejected real change in the world. Could the concept of permanent substance be preserved while still allowing for change? The atomists, hoping to account for the observable changes in the world, said yes.

Leucippus of Miletus (ca. fifth century BCE) and Democritus of Abdera of Trace (ca. 460–370 BCE) are credited with founding atomism. To them, atoms are perpetually moving, physically indivisible objects which are real and permanent. A similar view of atoms emerged in the East following a debate between two schools of Buddhist thought.

The Hinayana school believed in the reality of "dhammas." Dhammas are used to denote the essence of an event, whether the event is mental, psychic, or physical. Attributed to the dhammas are the reality and permanence of the Greek atoms. The Hinayanas taught that only a limited number of dhammas exist, but the exact number has been disputed within the Hinayana community.

By contrast, the Mahayana school of thought believed that dhammas were no more real or eternal than the human soul. They argued that everything in the observable world was flux. They did believe in an absolute, however. Their absolute could be reached by recognizing that the apparent permanence of an object was in fact only a relative property with no permanence. By sifting out all of the relative properties, the absolute would remain. Their word for the "absolute" is "shunyata," which is sometimes translated as void.

Greek atomists did not believe that atoms could be observed with human senses. Early atomists believed that atoms are eternally moving in a void; always combining and recombining to form the myriad of objects familiar to us in the observable world. Leucippus and Democritus espoused a view which rejected the influence of chance on the behavior of their atoms. Theirs was a deterministic view: a view which sees a direct cause for everything that happens. This mechanistic view of nature was accepted by the scientific community until the early part of the twentieth century when quantum mechanics first made its appearance. The most fundamental notions of the atomism of Leucippus and Democritus — permanence and indivisibility — played an important role in the development of modern science.

Table 13.1 summarizes ancient Greek philosophy of interest to us through the atomists. The atomists did not postulate a reason for the existence of motion of their atoms. They tried to account for the changing world without speculating on why the world was as it was. Their atomic theory

Table 13.1 Ancient Greek philosophy through the atomists.

Anaxagoras (500–428 BCE)	Primeval chaos ordered by cosmic mind
Heraclitus (535–475 BCE)	Change is the only reality
Parmenides (ca. 515 BCE)	Change is illusory
Democritus (ca. 460–370 BCE)	Atoms are permanent

did not address the question: why? Successors of the atomists, notably Plato and Aristotle, sought an answer to this question.

Plato

Plato (428–347 BCE) searched for the essence of objects. To him, specific things may change or pass away, but the Idea or Form embodied by a thing persists. Plato made this point clear by using a mode of expression common to his age, the dialogue. Plato credited his teacher Socrates (470–399 BCE) with answering the questions put forth in the following dialogue:

> **Q13.1.** "Then those who see the many beautiful, and who yet neither see absolute beauty, nor can follow and guide who points the way thither; who see the many just, and not absolute justice, and the like, — such persons may be said to have opinion but not knowledge?"
>
> "That is certain."
>
> "But those who see the absolute and eternal and immutable may be said to know, and not have opinion only?"
>
> "Neither can that be denied."
>
> [Stroll and Popkins, 1972, p. 35]

Socrates' views are summarized in Table 13.2. Our knowledge of Socrates' work is based on his student Plato's writings.

Plato believed that Ideas or Forms are unchanging and real. He further argued that Ideas can be known only by reason, not by sense experience. In other words, Plato and his adherents would accept arguments based on reason, such as mathematical proofs, but would reject observations because observations are merely sense experiences. This Platonic view is the antithesis of the Taoist belief espoused by Chuang Tzu: "The most extensive knowledge

Table 13.2 Socrates (470–399 BCE).

• Framed universal ideas, such as the idea of an immortal soul
• Soul: the center of our being and the locus of personality and intelligence
• True existence, if revealed to the soul at all, is revealed through thought, not by sense

does not necessarily know it [reality]; reasoning will not make men wise in it. The sages have decided against both these methods" [Capra, 1976, p. 101]. Taoism, which respects sensual experience, warns us against becoming too reliant on reason. Modern science moderates the extremes of Platonic and Taoist teachings by encouraging the interplay between reason (theory) and sense experience (experiment).

The Ideas or Forms of objects were Plato's raw material for answering the "how" of nature. The "why" of nature was provided by the source of reality of the Ideas. This source, the ultimate Idea, Plato called the Good: "Now, that which imparts truth to the known and the power of knowing to the knower is what I would have you term the idea of good, …" [Stroll and Popkins, 1972, p. 36]. Christian Platonism equates Plato's Good with God. The Good was the ultimate reality and the source of all that was knowable in the real world:

> **Q13.2.** … my opinion [Plato's opinion, as expressed by Socrates] is that in the real world of knowledge the idea of good appears last of all, and is seen only with an effort; and, when seen, is also inferred to be the universal author of all things beautiful and right, parent of light in this visible world, and the immediate source of reason and truth in the intellectual … [Stroll and Popkins, 1972, p. 41]

But the real world, the world of Ideas, was not the only world recognized by Plato. There was also the almost unreal world of specific things. To Plato, human beings occupy both worlds.

Similarities between the Hindu view of nature and the two worlds of Plato are striking. Hinduism teaches that the observable world is maya, meaning deceptively tricky. We must free ourselves from the illusory world (Plato's sensible world) before we can reach the real world (Plato's world of Ideas) of Brahman, the supreme reality.

A final point about creation should be made about Plato's metaphysics before we consider the metaphysics of Plato's student, Aristotle. Rather than creating the world from nothing, Plato believed God merely brought order to an already existing chaotic universe. Plato's belief is one side of an issue that is still unresolved: Has the substance of the universe always existed, or was the substance created *ex nihilo* (from nothing)?

Aristotle

Plato's student Aristotle (ca. 384–322 BCE) proposed an alternative to Plato's metaphysics summarized in Table 13.3. Aristotle relegated the most important tenet of Plato's metaphysics, the primacy of reason, to a secondary role. In its place, Aristotle substituted what Plato had explicitly rejected: sense experience.

Aristotle assumed that the world we experience is the real world. Unlike his predecessors — notably Parmenides, the Atomists, and Plato — Aristotle believed we could discover the features of nature in what we experience. To do so, we must recognize that all objects are continuously changing. From this flux, we must sift out what changes from that which remains the same. What is real is unchanging. Thus, Aristotle replaced the Atomists' atoms and Plato's Ideas with motionless objects as the basic building blocks of the universe. What, then, are these motionless objects?

Answering this question required the introduction of another postulate: the changing process is always directed toward some goal. The goal of every object, said Aristotle, was the attainment of rest and unchangeability which he termed the "prime mover": "We have shown that motion must be eternal and can never cease; so there must be some prime mover, whether singular or plural, that is eternal and not itself movable" [Stroll and Popkins, 1972, p. 218]. Aristotle's "prime mover" has become known as the "Unmoved Mover." The Prime Mover or Unmoved Mover is the motionless object as well as the Aristotelian God. It is a Being who does not act, yet attracts everything else just by its presence. Aristotle's concept of God was also the God of sainted Italian philosopher Thomas Aquinas. Table 13.4 presents a summary of Aristotle's metaphysics.

Two especially noteworthy features of Aristotle's metaphysics deserve highlighting. First, Aristotle espoused a teleological metaphysics, that is, a metaphysics with a purpose. This is evident in his requirement that changing

Table 13.3 Plato (428–347 BCE).

- The Idea or Form of an object is immutable
- Ideas are the source of reality
- Accepted *creatio ex nihilo*
- The Ultimate Idea is "Good"
- Accepted primacy of reason (ideas); rejected sense experience

Table 13.4 Aristotle (384–322 BCE).

• Accepts primacy of sense experience; reason is secondary • That which is real is unchanging • True understanding of an object requires knowledge of its "final cause," or the end for which it exists • Accepts Prime Mover: the being who does not act, yet attracts everything else by its presence

processes in nature should always be directed toward some goal which we have seen is the Unmoved Mover.

The second noteworthy feature stems from Aristotle's belief that every object seeks to be motionless and unchanging. His belief contrasts sharply with the first of the three postulates of mechanics set forth by Isaac Newton (1643–1727): "Every body continues in its state of rest, or of uniform motion in a right line, unless it is compelled to change that state by forces impressed upon it" [Wolff, 1965, p. 166]. The success of Newton's mechanics in describing the behavior of the observable world cast doubt upon Aristotle's metaphysics.

Post-Renaissance Metaphysics

Plato's metaphysics dominated the metaphysics of ancient times. Eventually, his views were superseded by Aristotle's teleological, or purposeful, metaphysics. Aristotle's teleological philosophy prevailed during the Middle Ages and up to the European Renaissance, a period when many cherished and long-held beliefs were challenged. The Renaissance ushered in a rediscovery of atomism and the stirring of intellectual freedom marked the birth of modern science. The dominance of ancient Greek philosophers on the contemporary philosophy of the Western world came to an end.

Plato's emphasis on reason and Aristotle's preference for sense experience were synthesized into a new approach to learning about nature, an approach that employed both reason and observation. Practitioners of this new approach were originally known as natural philosophers and have since evolved into what we now call scientists. Representative metaphysical ideas which have influenced the development of modern science are discussed here.

Descartes

Metaphysicists of historical renown abounded in the seventeenth century. They were heirs to many recent and significant discoveries. New continents were discovered. The earth was circumnavigated and found to be a small part of a vast universe. How did humanity fit into this new world? Was the concept of God necessary? If so, how did man relate to God? French philosopher Rene Descartes (1596–1650) was one of the first post-Renaissance metaphysicists to suggest a notable answer to these questions.

According to Descartes, the world was composed of three realities: God, mind, and matter. Descartes believed mind and matter were two different kinds of substance. Mind, said Descartes, is the substance that thinks. Matter, on the other hand, is a substance which has extension — it occupies a finite region of space — and can be known through its mathematical properties. Two points are raised by these statements. First, Descartes' belief in the distinguishability of mind and matter is the identifying characteristic of a dualist, that is, a person who believes in two kinds of substance. The second point stems from Descartes' reliance on mathematics to describe nature.

Various forms of mathematics, particularly Euclid's geometry, have existed since ancient times. Occasionally these ancient mathematical systems would be applied to practical problems. For example, the mathematician Apollonius of Perga (262–190 BCE) and astronomer Hipparchus of Nicaea (190–120 BCE) developed a complex geometric system based on the use of circles and the motion of circles on circles — so-called epicycles — to describe the motions of heavenly bodies. The Mayans of Mexico developed a highly sophisticated calendar system by the fifth century CE. Despite examples such as these, mathematics was not widely known or used to describe natural phenomena prior to the Renaissance.

Early post-Renaissance attempts to mathematically describe and predict natural phenomena included German astronomer Johannes Kepler's (1571–1630) formulation of the three laws of planetary motion and the development of Isaac Newton's mechanics. Their efforts influenced philosophical thought because they provided reasonable explanations and accurate predictions of some natural phenomena. Descartes readily incorporated mathematics into his metaphysics.

What is mathematics? In Chapter 10, we discussed the value of logic systems — or "rules of the game" — in two areas: developing a consensus among individuals and making predictions. By changing the rules of the logic system, we can construct as many different logic systems as we want. The same is true of mathematics. In fact, we can regard mathematics as a compact manifestation of logic systems.

Different logic systems correspond to different mathematical structures which, in turn, are simply different sets of rules we have agreed to abide by. The discovery that mathematics can describe some natural phenomena may be interpreted to mean that nature, or at least some aspects of nature, behave in a logical fashion. This discovery was familiar to ancient philosophers and was a cornerstone of Plato's metaphysics, namely his preference for reasoning. Plato's dismissal of sense experience, however, hindered his efforts to develop a logic system or set of postulates that were as useful as, say, Isaac Newton's in accurately describing nature.

Returning now to Descartes' metaphysics, Descartes argued that God was necessary as the creator and preserver of mind and matter. Unlike mind and matter, however, God must be self-caused. This assertion by Descartes rests on the assumption that causality exists in nature.

Descartes' extended objects — his matter — were allowed to interact with one another. The extended objects and their interactions operated like a machine which was activated and maintained by God. Descartes advocated the mechanistic view of nature: unless God changed natural laws, the behavior of nature in the future could be completely determined from its past behavior if we understood and could apply God's natural laws. Descartes' metaphysics did not support the concept of free will in nature.

Descartes believed his first substance (the nonphysical mind) and his second substance (physical matter) were connected in the pineal gland. The pineal gland is a small, usually conical appendage of the brain of all cranial vertebrates, including humans. It has a shape similar to the human eye and is located at the base of the human brain. The pineal gland produces the hormone melatonin, which is useful for promoting sleep. Some people have suggested that the pineal gland may be the point of connection between the body and the soul.

In addition to the two realities of mind and matter, Descartes taught of a third reality: God. Descartes argued that the truth of all knowledge, which is known only through the mental world, is guaranteed by God.

> **Q13.3.** … every clear and distinct conception is doubtless something, and as such cannot owe its origin to nothing, but must of necessity have God for its author — God, I say, who as supremely perfect, cannot, without a contradiction, be the cause of any error; and consequently, it is necessary to conclude that every such conception or judgement is true. [Stroll and Popkins, 1972, p. 69]

Thus, the three realities of Descartes' metaphysics — God, mind, and matter — were inextricably linked. Not all of Descartes' contemporaries agreed, however, with the uniqueness of the relationships Descartes attributed to his three realities.

Hobbes

English philosopher Thomas Hobbes (1588–1609) said there were not two distinct substances in the world, as dualists such as Descartes advocated, but only one substance. Hobbes believed the only substance in the world was matter, a special case of which was Descartes' mind. To Hobbes, then, the world was composed only of God and matter. The multiplicity of forms people envisaged, said Hobbes, is the result of external objects acting on our sensory systems:

> **Q13.4.** … [thoughts] are every one a *representation* or **appearance** of some quality or other accident of a body which is commonly called an *object*. Which object works on the eyes, ears, and other parts of a man's body, and by diversity of working produces diversity of appearances. [Stroll and Popkins, 1972, p. 222]

Hobbes espoused a monistic view: a belief in only one ultimate substance in nature besides God. For Hobbes, the problem of understanding all mental processes reduces to the problem of determining how the brain responds to signals it receives from receptors of the nervous system. The receptors, in turn, have been stimulated by external objects. The process is thought to be mechanistic.

Hobbes' monism, or monistic view, is representative of a materialistic, deterministic view of nature. He applied his deterministic concepts to a wide range of subjects, including politics and psychology. The application of determinism to psychology provided a basis for psychological theories such as behaviorism.

Spinoza

Dutch philosopher Baruch Spinoza (1632–1677) considered reality to be a single substance, and that substance was God: "God, or substance, consisting of infinite attributes, of which each expresses eternal and infinite essentiality, necessarily exists" [Stroll and Popkins, 1972, p. 248]. To Spinoza, there could be no substance other than God: "Besides God no substance can be granted or conceived" [*Ibid.*, p. 250]. Thus, Spinoza was neither a monist like Hobbes nor a dualist like Descartes. Spinoza's views were so controversial in his time, particularly with the Catholic Church, of which he was a member, that he was excommunicated for advocating atheism.

Spinoza did not make a distinction between the creator and its creations. To Spinoza, all things were merely different forms of one substance. This view is much like the pantheistic view of Hinduism, as twentieth-century physicist Fritjof Capra (1939–) relates:

> **Q13.5.** The basis of ... all Hinduism, is the idea that the multitude of things and events around us are but different manifestations of the same ultimate reality. This reality, called BRAHMAN, is the unifying concept which gives Hinduism its essentially monotheistic character in spite of the worship of numerous gods and goddesses. [Capra, 1976, p. 77]

Unlike Hinduism, however, Spinoza considered worship worthless because his God was totally impersonal: "God acts solely by the laws of his own nature, and is not constrained by anyone" [Stroll and Popkins, 1972,

Table 13.5 Metaphysical arguments: Descartes, Hobbes, and Spinoza.

Descartes (1596–1650 CE)	• Proponent of God, mind and matter (dualist) • God: creator and preserver of mind and matter • Truth of all knowledge guaranteed by God • Dualist: mind and matter are distinct • Mechanistic (deterministic) view of nature
Hobbes (1588–1679 CE)	• Proponent of God and matter (monist) • Mental processes are subset of matter • Deterministic view
Spinoza (1632–1677 CE)	• Pantheistic view: all things are different forms of one substance • One substance: God

p. 251]. Spinoza's world was comprised of one substance which could be understood rationally. He believed in the inseparability of God and nature.

If everything in nature was just a different manifestation of God, Spinoza argued, then the more traditional explanations, including those advocated by the powerful Catholic Church, were unnecessary: "... all the explanations commonly given of nature are mere modes of imagining, and do not indicate the true nature of anything, but only the constitution of the imagination" [Stroll and Popkins, 1972, p. 257]. Although Spinoza believed in a god, it was not the God of his Catholic Church. Spinoza was excommunicated and became an outcast of his native intellectual community. Spinoza earned his living as a lens grinder rather than as a philosopher. Although this turn of events complicated his life, it played a useful role in history because it demonstrated that a scholar need not be part of the intellectual establishment as a precondition to contributing to humanity's store of knowledge. This was an important lesson for one of the most influential scientists of all time: Albert Einstein (1879–1955).[2]

The metaphysical arguments of Descartes, Hobbes, and Spinoza are summarized in Table 13.5.

Leibniz

A metaphysics which differs significantly from those presented previously, yet has features resembling those of some modern physical theories, is the metaphysics propounded by the German mathematician and philosopher Baron Gottfried Wilhelm von Leibniz (1646–1716).

Leibniz introduced the monad as his basic natural entity: "The Monad, of which we will speak here, is nothing else than a simple substance, which goes to make up composites; by simple, we mean without parts" [Stroll and Popkins, 1972, p. 259]. The monad has no extension — it has no volume at all — and does not interact with other monads. It corresponds roughly to an elementary particle except its motion is completely independent of the motion of all other monads. The independence of Leibniz's monads is the most unusual aspect of his metaphysics. If monads are not allowed to interact with one another, then how do we explain what appear to be causal relationships?

Every monad, says Leibniz, moves in a pre-established harmony with respect to every other. As an illustration, suppose I am near the door when the doorbell rings. If I hear the doorbell ring, was the sound I heard caused by the ringing doorbell? Leibniz would say the ringing doorbell and the sound of the doorbell represent the harmonious interplay of noninteracting monads.

If monads exist and possess the characteristics Leibniz assigned to them, an obvious question arises: who or what establishes and orchestrates their harmonious dance? Leibniz answers:

> **Q13.6.** In the case of simple substances, the influence which one monad has upon another is only ideal. It can have its effect only through the mediation of God, in so far as in the ideas of God each monad can rightly demand that God, in regulating the others from the beginning of things, should have regarded it also. For since one created monad cannot have a physical influence upon the inner being of another, it is only through the primal regulation that one can have dependence upon another. [Stroll and Popkins, 1972, p. 264]

Thus, Leibniz identifies God as the orchestrator. Furthermore, the harmony exists without contradiction because God — who is all-wise, all-good, and all-knowing — was able to select the most harmonious of all possible worlds. This world, says Leibniz, is the best of all possible worlds:

"Through this means [selection] has been obtained the greatest possible variety, together with the greatest order that may be: that is to say, through this means has been obtained the greatest possible perfection" [Stroll and Popkins, 1972, p. 265].

Many of Leibniz's contemporaries had as much trouble believing their world was the best possible world as many of us do about our world. It should come as no surprise that Leibniz has been considered an optimist. Indeed, the French philosopher and author, Francois Arouet de Voltaire (1694–1778), ridiculed Leibniz's optimistic view of the world — as embodied by the character Dr. Pangloss — in his book *Candide*. Rather than join in the ridicule, let us see if any of Leibniz's ideas can aid us in our search today.

Inherent in Leibniz's metaphysics is the possibility of the existence of worlds other than our own. This notion is present in some modern views of nature. For example, the Many Worlds Interpretation (MWI) proposed in 1957 by American physicist Hugh Everett III (1930–1982) hypothesized that all possible outcomes of a quantum measurement are physically realized in different equally real "worlds" or universes [Byrne, 2007]. The set of "worlds" or universes from the MWI are known as the multiverse.

Hume

The 17th- and eighteenth-century philosophers John Locke (1632–1704), George Berkeley (1685–1753), and especially David Hume (1711–1776), expressed skepticism about the value of metaphysics. They preferred an empirical theory which the Scotsman Hume developed most completely.

According to Hume, the existence of substance in general, and being in particular, could not be proved except by recourse to empirical arguments: "The existence ... of any being can only be proved by arguments from its cause or its effect; and these arguments are founded entirely on sense experience" [Stroll and Popkins, 1972, p. 193]. Hume expressed a strong reliance on sense experience. He did not accept Descartes' contention that the truth of all knowledge could be known only through the mental world. Hume rejected cogitative proof of the existence of Self. Of greater interest here, however, is Hume's view of causality.

Causality expresses a relationship between two events. We say event A caused event B if the appearance of event A is always accompanied by the appearance of event B. The relationship between events A and B is said to be a causal relationship. As an illustration, suppose your phone rings and no one else is nearby. You will probably answer the phone. We are associating the ringing of the phone (event A) with the answering of the phone (event B). A causal relationship exists between the ringing phone and you're answering it. Sometimes, however, you may not answer the ringing phone. Perhaps you are in the shower or you are in a rebellious mood. The causal relationship is broken when the phone is not answered. Is this significant?

A determinist would say no. He might support his answer by claiming you are merely responding to a more powerful causal influence. Hume, by contrast, would consider the breaking of one causal relationship a significant piece of evidence in support of an underlying reality: all causal relationships can be broken. To Hume, causality is nothing but an association we have inferred between two or more events. If we always observe event B following event A, we are inclined to say event A causes event B. This claim is a hypothesis based on historical observations; it is not a fact based on abstract and infallible deliberations. As Hume says, "… all our reasonings concerning causes and effects are derived from nothing but custom; and that belief is more properly an act of the sensitive, than of the cogitative part of our nature" [Russell, 1972, p. 671]. Does this mean no causal relationships exist?

We have already noted an example of a causal relationship: the ringing telephone. The question is not whether causal relationships exist, but do any unbreakable causal relationships exist? In the past, especially from the days of Isaac Newton to the beginning of the twentieth century, it was thought that inviolable causal relationships did exist. These relationships were considered the basis of the mechanistic world. So successful was Newton's theory of mechanics that the hypothesis of inviolable causality became a virtually undisputed principle or "law." Only since the early 1900s have scientists restored causality as a postulate which might not be correct.

Kant

Historically, the ascension of a specific metaphysics was often followed by attempts to refute or otherwise diminish that metaphysics. Hume's view was not an exception. Unlike Hume, who doubted the "truth" of metaphysical knowledge, German philosopher Immanuel Kant (1724–1804) sought to show how people could acquire metaphysical knowledge.

Kant sought knowledge which was both necessarily and universally true. Some of this knowledge, he said, should apply to all possible experiences. Much of Kant's metaphysics was an attempt to correct problems he saw in Hume's metaphysics. One of Kant's major concerns was Hume's views on the basis of necessary (indisputable) knowledge:

> **Q13.7.** The question was not whether the concept of cause was right, useful, and even indispensable for our knowledge, for this Hume never doubted; but whether that concept could be thought by reason *a priori*, and consequently whether it possessed an inner truth, independent of all experience, implying a wider application than merely to the objects of experience. This was Hume's problem. It was a question concerning the *origin*, not concerning the *indispensable need* of the concept. [Stroll and Popkins, 1972, p. 272]

What was the origin of necessary knowledge? The source of necessary knowledge, Kant suggested, was not the observed events, but the person observing the events:

> **Q13.8.** I therefore first tried whether Hume's objection [to *a priori* causality] could not be put into a general form, and soon found that the concept of the connexion of cause and effect was by no means the only idea by which the understanding thinks the connexion of this *a priori*, but rather that metaphysics consists altogether of such connexions. I sought to ascertain their number, and when I had satisfactorily succeeded in this by starting from a single principle, I proceeded to the deduction of these concepts, which I was certain were not deduced from experience, as Hume had apprehended, but sprang from the pure understanding. [Stroll and Popkins, 1972, p. 273]

In more modern terms, causal relations do not necessarily exist between events but are imposed by the person observing the events. The observer

Table 13.6 **Metaphysical arguments: Leibniz, Hume, and Kant.**

Leibniz (1646–1716)	• Matter consists of monads • Monads do not interact; they move in pre-established harmony • God orchestrates the behavior of monads • God is all-wise, all-good, all-knowing: therefore the world is best of all possible worlds
Hume (1711–1776)	• Empiricist: relies on sense experience • Causal relations are empirical, not immutable
Kant (1724–1804)	• Scientific knowledge depends on the observed events and the observer • Metaphysical knowledge cannot be discovered by reason alone

creates causal relations as part of a more general classification scheme. By categorizing sensory input, we can describe it with greater ease and perhaps gain an understanding of the observations we have made. This cataloging process accounts for the necessarily true knowledge — the raw material — of science. Thus, concluded Kant, scientific knowledge is a product not only of the events observed but of the observer as well.

Table 13.6 summarizes the metaphysical arguments of Leibniz, Hume, and Kant.

❁ ❁ ❁

Metaphysical ideas have ranged from the denial of substance to the belief in substance; from "proofs" of God's existence to the excommunication of a man for his alleged atheism; and from the acceptance to the rejection of causality. We have ended with Kant.

Kant did not think metaphysical knowledge could be discovered through reason alone. He said we could not catalog or think about substance relevant to metaphysics without making it amenable to experiment. Once a substance becomes subject to experimental testing, it becomes empirical and stops being metaphysical. Was Kant right? What is the meaning of metaphysical knowledge? For that matter, what do we mean by the word "knowledge"? How do we know when we know?

Chapter 14

Knowledge and Science

The scientific worldview presented in Part 1 is based on knowledge acquired by the scientific method. We reported evidence for the scientific worldview without questioning the veracity of the method used to acquire the evidence. In this chapter, we review key concepts of epistemology,[1] the branch of philosophy concerned with determining what knowledge is and how it can be obtained. We then examine how knowledge is acquired using the scientific method.

Rationalism

Our search for the meaning of knowledge begins with rationalism. Rationalists are adherents to rationalism who have based their philosophies primarily on reason. Some of the earliest rationalists had a dim view of humanity's ability to acquire knowledge that was both true and irrefutable. Prominent among these early skeptics was the Greek philosopher Protagoras (ca. 490–420 BCE). Protagoras was a Sophist: a teacher of young men somewhat akin to the modern professor. As a rule, Sophists pursued truth. Often their pursuit led them to seriously doubt the existence of indisputable knowledge. Man, so it seemed to Sophists, could only espouse beliefs or opinions. If this view is correct, it would seem to be a useless exercise to pursue philosophical truths. Even during the time of Sophist preeminence in ancient Greece, the Sophist view was challenged.

One of the most prominent challengers was the Greek philosopher Socrates (ca. 470–399 BCE), a contemporary of Protagoras. Socrates believed that people should be able to find real knowledge, that is, something we might call truth. To Socrates, truth was necessary as a justification of our behavior. Without truth, what would we have as a basis for our actions? Yet if we can know truth, what truth is there for us to know? In the *Apology*,

Socrates gave his answer to this question while responding to the inquisition of an Athenian court:

> **Q14.1.** … the truth is, O men of Athens, that God only is wise; and by his answer he intends to show that the wisdom of men is worth little or nothing; he is not speaking of Socrates, he is only using my name by way of illustration, as if he said, He, O men, is the wisest, who, like Socrates, knows that his wisdom is in truth worth nothing. [Socrates as reported by Plato; see Stroll and Popkins, 1972, p. 11]

Unfortunately, his humility did not spare Socrates from punishment by death for his "subversive activities" against Athens, activities which sought to reveal the pomposity of the ruling class. Athenians were able to execute Socrates, but they were unable to destroy his ideas. These ideas lived on through the writings of his students, most notably Plato (428–347 BCE).

Plato tried to show what could be known and how. As we have seen, Plato believed the only reality was embodied in the Idea or Form of an object. He said the knowledge of the Ideas of objects resides in our memories. It is up to us to exercise our power to learn, a power Plato says is present in our souls:

> **Q14.2.** … if I am right, certain professors of education must be wrong when they say that they can put a knowledge into the soul which was not there before, like sight into blind eyes. … our argument shows that the power and capacity of learning exists in the soul already … [Stroll and Popkins,1972, p. 41]

Plato's power to learn is analogous to recollection. According to this view, we are born with all the knowledge we will ever possess. Then, employing the intellectual activity known as recall, we retrieve a fraction of this knowledge. Knowledge gained in this way is supposed to be permanent and applicable to the real world of Ideas.

Two millennia after the period during which the Sophists thrived, their skeptical theory of knowledge was still being disputed. This time their challenger was French philosopher René Descartes (1596–1650). Descartes challenged the Sophists' skeptical views by carrying skepticism to its extreme and thereby exhibiting its flaws. Descartes began by accepting the Sophists' basic premise in its most extreme form: he rejected every thought or piece of information which might be false. Nothing remained except his ability to think. This ability, he surmised, must therefore be real. But if Descartes could think, then he, too, must be real, thus he concluded:

> **Q14.3.** … I had the persuasion that there was absolutely nothing in the world, that there was no sky and no earth, neither minds nor bodies; was I not, therefore, at the same time, persuaded that I did not exist? Far from it; I assuredly existed, since I was persuaded. But there is I know not what being, who is possessed at once of the highest power and deepest cunning, who is constantly employing all his ingenuity in deceiving me. Doubtless, then, I exist, since I am deceived; … [Stroll and Popkins, 1972, p. 58]

Descartes's realization has been succinctly summarized by the phrase "I think, therefore I am."

With this realization, Descartes no longer doubted his ability, at least in principle, to know truth. Descartes proceeded to suggest a means for acquiring truth: "… for I will assuredly reach truth if I only fix my attention sufficiently on all the things I conceive perfectly … [Stroll and Popkins, 1972, p. 69]."

Descartes believed that he could form three perfectly conceived ideas: God, mind, and the essence of matter. Descartes, as we saw in the previous chapter, was a dualist. He accepted the idea that mind and matter were distinct objects. Descartes argued that we can discover truth and God by reasoning:

> **Q14.4.** And the whole force of the argument of which I have availed myself to establish the existence of God, consists in this, that I perceive I could not possibly be of such a nature as I am, and yet have in my mind the idea of a God, if God did not in reality exist, … [Stroll and Popkins, 1972, p. 68]

To Descartes, perfectly conceived objects such as God must be clear and distinct. Does the notion of perfect conception exclude sense information? Descartes and his fellow rationalists would say yes while simultaneously reminding us of the unreliability of sense information. There are those, however, who disagree with the rationalists.

Eastern Empiricism

Unlike rationalists, empiricists rely on sense information. Empirical theories are fundamental to modern science and have been advocated by people throughout the ages. The Taoist, Chuang Tzu (ca. 369–286 BCE), warned us to distrust those who rely only on reason and persuasion [Capra, 1976, p. 101]: "A dog is not reckoned good because he barks well, and a man is not reckoned wise because he speaks skillfully." Many in the East and the West have grown to accept the information they receive through their senses. In this section, we discuss two representative examples of Eastern empiricism, while in the next section our attention turns to Western empiricism.

Taoism, as we have seen, is distrustful of knowledge derived by reason or rational argument. Taoists prefer intuitive knowledge to rational knowledge because they believe intuitive knowledge derives from an intimate relationship between themselves and the rest of nature. Yet, strictly speaking, Taoists do not subscribe to the notion that absolute knowledge can be obtained. In fact, Taoists believe knowledge is never absolute: the validity of knowledge depends on the perceiver. The Taoist view is symbolized by the Yin-Yang (Figure 14.1).

The Yin and the Yang represent polar opposites. Feminine traits, intuitive knowledge, passive responses, and evil behavior are typically represented by the Yin; whereas Yang depicts their opposites, namely masculine traits, rational knowledge, active responses, and good behavior. As Taoist K.K. Yzu tells us [Capra, 1976, p. 97], "The Yang returns cyclically to its beginning; the Yin attains its maximum and gives place to the Yang."

Figure 14.1. Yin-Yang.

The dark dot in the Yang and the light dot in the Yin are symbolic of the Chinese concept that the Yin and the Yang always contain some semblance of their opposite, even when they are at their peaks. How can we know good, a Taoist may ask, if we have no evil to compare it with? Thus, a shred of evil lurks in the very best, and a shred of good sparkles in the very worst. Seemingly distinguishable and independent concepts, such as good and evil, are relative concepts. The pairs of concepts symbolized by the Yin-Yang are complementary.

Another representative of Eastern empiricism is typified by Zen Buddhism. Zen Buddhism, or simply Zen, is the intuitive school of Buddhism. Zen depends largely on enlightenment or intuition for the attainment of knowledge. Zen Buddhists believe truth can never be completely expressed in words. Truth, they contend, must be experienced. Zen, perhaps more than other religions, stresses an awareness of the limitations of language in constructing an accurate representation of the real world. With such serious reservations about our best means of communication — language — how does Zen survive from one generation to another? Huston Smith (1919–2016) tells us:

> **Q14.5.** Zen has rested its survival on the transmission of a specific state of awareness directly from mind to mind, like flame passed from candle to candle, or water poured from bucket to bucket. [Smith, 1965, p. 145]

The "state of awareness" characteristic of Zen is transmitted from one mind to another by the training of Zen students at the feet of Zen masters. The master poses a long series of riddles, known as koans, to the student. The student must answer the riddles to the master's satisfaction. Some exemplary koans include:

> **Q14.6.** What was the appearance of your face before your ancestors were born? [Smith, 1965, p. 146]

and

> **Q14.7.** We are all familiar with the sound of two hands clapping. What is the sound of one hand? [Smith, 1965, p. 146]

Koans appear to be absurdities to minds accustomed to reason. This is one of the effects they are intended to have, as Fritjof Capra explains:

> **Q14.8.** Koans are carefully devised nonsensical riddles which are meant to make the student of Zen realize the limitations of logic and reasoning in the most dramatic way. [Capra, 1976, p. 35]

Only by shedding conventional logic can the student hope to understand koans. When, and if, understanding is finally achieved, it arrives as an intuitive experience called satori. The student has transcended the opposites of nature and is now one with the single Buddhist reality known as Nirvana.

Western Empiricism

Unlike Eastern empiricism, which relies on the "internal" sense experience of intuition, Western empiricism depends on the five "external" sense experiences: sight, smell, touch, taste, and hearing. One of the first protagonists of Western empiricism was English philosopher John Locke (1632–1704).

Locke argued against Plato's belief that people are born with all the knowledge they will ever have. Instead of Plato's innate ideas, Locke said that we are born with only a few basic instincts, and we must rely on our experiences to shape our lives. His supposition has become known in psychology as the "blank slate" hypothesis. Ideas appear in the mind, said Locke, either as a direct result of sensory experience or indirectly from thinking about experience:

> **Q14.9.** The senses at first let in *particular* ideas, and furnish the yet empty cabinet, and the mind by degrees growing familiar with some of them, they are lodged in the memory, and names got to them. Afterwards, the mind proceeding further, abstracts them, and by degrees learns the use of general names. In this manner the mind comes to be furnished with ideas and language, ... [Stroll and Popkins, 1972, p. 83]

Given these ideas as basic concepts, how did Locke determine knowledge?

Three methods of determining knowledge were recognized by Locke: intuition, demonstration, and sensation. According to Locke, certain knowledge, such as our existence, was based on intuition:

> **Q14.10.** … sometimes the mind perceives the agreement or disagreement of two ideas *immediately by themselves*, without the intervention of any other: *and* this I think we may call *intuitive knowledge.* … It is on this intuition that depends all the certainty and evidence of all our knowledge; … [Stroll and Popkins, 1972, p. 101]

Influenced by Descartes' metaphysics, Locke claimed the knowledge of God's existence could be established by demonstration:

> **Q14.11.** Those intervening ideas, which serve to show the agreement of any two others, are called *proofs*; and where the agreement and disagreement is by this means plainly and clearly perceived, it is called *demonstration* … [Stroll and Popkins, 1972, p. 102]

Finally, Locke said we could know of the existence of certain external objects when we were sensing them:

> **Q14.12.** … we may add to the two former sorts of knowledge this also, of the existence of particular external objects, by that perception and consciousness we have of the actual entrance of ideas from them, and allow these three degrees of knowledge, namely *intuitive*, *demonstrative*, and *sensitive* … [Stroll and Popkins, 1972, p. 106]

Locke's notion of a material world existing outside of our minds was not well received by a successor and fellow Englishman, George Berkeley (1685–1753). Berkeley thought it was unreasonable to believe in the existence of objects external to ourselves. He said we could only know the idea of an object rather than the object itself. Furthermore, Berkeley reminded us of the necessity of the mind for the existence of ideas. He said that "… no idea can exist without the mind" [Stroll and Popkins, 1972, p. 142]. Does this imply external objects will cease to exist when we are no longer thinking of them? Berkeley answered "no" because those external objects, even though they are not perceived by us, are perceived by God. They, therefore, exist from God's point of view.

David Hume (1711–1776) took a more skeptical view of our ability to acquire knowledge than did Locke and Berkeley. All we can do, said Hume, is describe our perceptions. Our "knowledge" is limited to these descriptions. If Hume's view is correct, what does it imply about those relationships between external objects which have been surmised from an analysis of our perceptions? More particularly, does Hume's view preserve causality?

Causal reasoning is based on what we expect to happen in a given situation as a result of our experience, either direct or shared, with other identical or similar situations. As Hume pointed out, the apparent causal connection associated with a given sequence of events is not guaranteed. To him, causality is nothing more than the observation that sequences of certain observed events have always occurred in the same order in the past. We do not "know" event A was the cause of a succeeding event B; we only "know" from our prior experience that A has always preceded B. As Hume relates:

> **Q14.13.** While we cannot give a satisfactory reason, why we believe, after a thousand experiments, that a stone will fall, or fire burn; can we ever satisfy ourselves concerning any determination, which we may form, with regard to the origin of worlds, and the situation of nature, from, and to eternity? [Stroll and Popkins, 1972, p. 192]

From a modern perspective, Hume is asking an important question: how do we know the laws of nature will continue unchanged into the future? The answer is simple: we don't know! It is a matter of faith on our part that the nature we have studied in the past will not become radically and arbitrarily different in the future. If the laws of nature inexplicably change, perhaps by the whim of an omnipotent deity, then the combined efforts of mankind could well be for naught.

Should we adopt the rationalists' theory of knowledge, or should we choose an empirical view? How does modern science determine knowledge? In the following section, we discuss the scientific method by considering it in the context of normal science and scientific revolutions.

Normal Science and Scientific Revolutions

Karl Popper (1902–1994), an Austrian and British philosopher of science, suggested that science is incapable of discovering truths. He argued that science gains knowledge by falsifying hypotheses rather than by verifying them. Scientific statements that are not falsified, says Popper, should not be considered verified or proved; they should be considered corroborated. In this sense, scientific statements are hypotheses that are useful in explaining historic phenomena and have been reliable in predicting future behavior. According to Popper, scientific knowledge must be falsifiable. His view of scientific knowledge was challenged in 1962 by American Thomas Kuhn (1922–1996) in his landmark work *The Structure of Scientific Revolutions* [Kuhn, 1970].

Kuhn introduced the concept of paradigm into the discussion of scientific revolutions. The paradigm is the current worldview in mainstream, or normal, science. The paradigm is our gestalt, or how we perceive the environment. Anomalies, or difficulties with the paradigm, can lead to a crisis in the practice of normal science. Resolution of the crisis can require a change in perception or gestalt shift.

Gestalt shifts require a transformation of our perception. Figure 14.2 entitled "My Wife and My Mother-in-Law" was drawn by cartoonist W.E. Hill in 1915. It illustrates the concept of gestalt shift. The image shows both an old lady and a young woman, depending on your perspective. Some people may have difficulty making the gestalt shift.

The new paradigm resulting from the change in perception can encounter resistance by proponents of the old paradigm, but the new paradigm gains acceptance as normal science when it resolves enough anomalies of the old paradigm to receive support from the mainstream of the scientific community. Acceptance brings closure to the revolution and establishes a new paradigm of normal science. Stages of scientific revolution in Kuhn's theory are displayed in Figure 14.3.

Figure 14.2. Can you make the gestalt shift?

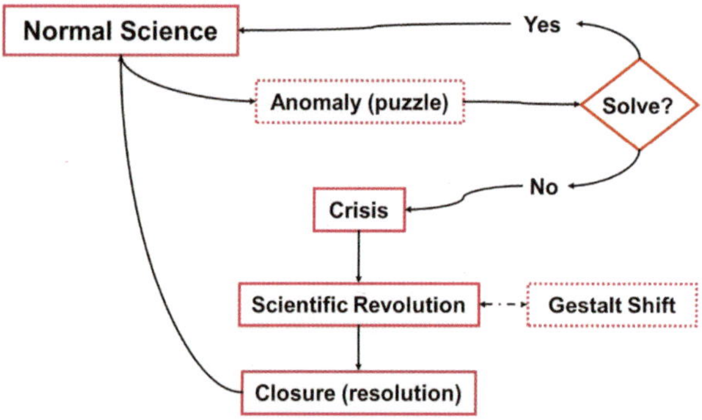

Figure 14.3. Kuhn's theory of scientific revolutions.

The birth and growth of competing paradigms are usually due to a continuing failure on the part of the paradigm of normal science to explain important anomalies. An anomaly is a persistent problem within the mainstream paradigm. Proponents of the paradigm of normal science view anomalies as puzzles to be solved within the context of normal science. Challengers of the paradigm of normal science view anomalies as evidence of a crisis in normal science. If the old paradigm is replaced by a new paradigm in which the anomalies are resolved, the period of crisis is closed, and the new paradigm becomes the new normal science as in Figure 14.3.

During periods of crisis, says Kuhn, it is not possible to rationally discuss conceptual differences or compare experimental results because competing paradigms may not overlap sufficiently for a common basis of communication to exist. Supporters of the old and the new paradigms may not be speaking the same language. Although the forms of the words they use are the same, the meanings have changed. Kuhn says that the competing paradigms are incommensurable. How, then, does an old paradigm get replaced?

Kuhn argued that the replacement is not achieved by logical falsification, but by a variety of causes, some of which are irrational. Proponents of the old paradigm may be converted to the new when the new paradigm is successful in resolving persistent anomalies associated with the old paradigm. Others may be converted because of the stature of their colleagues who have switched to the new paradigm or who helped develop the new paradigm.

Falsification

Kuhn's work was a pioneering effort to discover the mechanisms behind major changes in humanity's understanding of nature through science. His work contrasts with that of Popper. Attempts to buttress Popper's falsification concepts have been made since the appearance of Kuhn's book. A notable model of scientific revolutions that employs an updated notion of falsification was proposed by Imre Lakatos (1922–1974).

Lakatos says scientists compare research programs instead of paradigms. According to Lakatos, a research program contains a hard core of basic assumptions. It includes perceptions of nature and its various constituents. The hard core of a research program also embodies unfalsified, widely accepted theories. In addition to a hard core, the research program has a heuristic part. The heuristic part consists of auxiliary hypotheses that help define the research to be done. From Lakatos' perspective, normal science exists as long as its research program can be extended to new areas, and testable predictions can be made. A research program becomes suspect if it must continually adopt new auxiliary hypotheses to explain persistent anomalies, or incorporate unexpected observations arising from competing research programs. The healthy research program is dubbed progressive, while the embattled research program is viewed as stagnating or regressive.

Many aspects of Lakatos' model have counterparts in Kuhn's model. Kuhn's paradigm, for example, encompasses much of what Lakatos calls the hard-core part of the research program, as well as portions of the heuristic part. The feature that distinguishes the two models is Lakatos' assertion that scientists can reject one research program in favor of another as a result of rational considerations or decisive experimental tests.

If an individual can be convinced of the validity of a competing paradigm or research program by rational means (such as a decisive experiment), then that person fits the framework of Lakatos' model and represents the ideal of a practicing scientist. On the other hand, if an individual cannot be reasoned with, or is converted to a new paradigm for irrational or scientifically irrelevant reasons (such as a loss of funding because the person does not conform), that person fits Kuhn's framework. One would hope that Lakatos' model would be correct. Unfortunately, my experience supports Kuhn's model of how some people behave.

Personal Anecdote: Dissemination of Nonconforming Ideas

A fact of life when working outside of normal science is the difficulty of getting a fair hearing from people who are too busy within normal science to also learn enough about a nonconforming theory to judge its value. This makes it difficult to publish in mainstream journals and to secure funding from such mainstream sources as government agencies. A way to get around this problem is to publish in journals that encourage innovation. One such journal is *Foundations of Physics.* In addition to publishing in special purpose journals, scientists can start new organizations and hold conferences to encourage the development of nonconforming theories and to facilitate the dissemination of ideas. That was what a group of us did when we formed the International Association for Relativistic Dynamics (IARD) in 1998 and published the refereed proceedings of our meetings in *Foundations of Physics.*

Table 14.1 **Historical examples of competing paradigms.**

Old Paradigm	New Paradigm
Phlogiston	Oxygen
Absolute time	Relative time
Corpuscles	Waves (and then wave-particle duality)
Geocentrism	Heliocentrism
Caloric	Heat
Continuous energy	Discrete energy
Action at a distance	Local interactions
Aether (speed of light not constant)	Vacuum (speed of light constant)
Gravitational force	Space-time geometry

Examples of Scientific Revolutions

Scientific revolutions occur when a new paradigm dethrones the substantial infrastructure supporting the mainstream paradigm. Historically, scientific revolutions do not occur easily, but revolutions in science have happened. They are characterized by the existence of competing paradigms. Table 14.1 lists several historical examples.

We do not know which model of scientific revolutions is correct. It is worth noting that normal science is built upon a set of beliefs (a paradigm in Kuhn's model) that may not be permanent. Scientific revolutions are evidence that the favored views of the scientific establishment can be replaced with a new set of equally favored beliefs by succeeding generations. It is important to remember this point when considering two competing paradigms, especially when one of the paradigms has the authority of the scientific establishment behind it. The validity of a scientific paradigm should not depend on the authority of the scientific establishment. Indeed, beliefs based on authority are not scientific, they are dogmatic. We need to think for ourselves.

Is Science Immutable?

Has the scientific method made it possible for humanity to discover absolute truths about nature; or should we believe that everything we learn about nature is transitory? Are scientific beliefs unchanging over time, or have they changed with time? Science historians have documented many significant changes in scientific beliefs. The historical reality of scientific revolutions demonstrates that scientific beliefs can change.

Human understanding of cosmological phenomena has changed over the centuries, and even over the past few decades. Major conceptual changes include moving the center of the Solar System from the Earth to the Sun and replacing Newton's absolute time and space with Einstein's relative time and space. Recent observations of the distribution of galaxies have shown that we have only observed a small percentage of mass and energy in the universe; the remaining dark mass and dark energy have yet to be detected. Are dark mass and dark energy anomalies that presage the coming of a new scientific revolution? What changes will have to be made to our current ideas about cosmology? These factors should be considered when answering the question: does god exist?

Modern cosmological models are based on the notion that the large-scale universe we see today would have once existed on the subatomic scale. This means that our understanding of cosmology is linked to our understanding of phenomena on the subatomic scale. Prior to 1900, scientists were able to use the classical physics of Newton, Maxwell, and many others to explain virtually all known experimental measurements. A few experimental anomalies motivated the reevaluation of many previously unchallenged concepts. The constancy of the speed of light in vacuum, for example, was understood only when Einstein rejected Newton's view of absolute and independent space and time. Another experimental observation, the radiation emitted by heated black bodies, spawned a scientific revolution on the subatomic scale.

Classical physics has been replaced in the scientific worldview by a scientific revolution on the subatomic scale. The new quantum theory is based on chance. The demise of classical determinism, including causality, as an indisputable law was recognized by the founders of quantum theory. The replacement of classical physics by quantum theory once again shows that the ideas of science can change and, when they do, can significantly alter what we thought we knew about nature.

Personal Observation: Work in Progress?

I have studied the question of god's existence since I was a teenager in the 1960s. I wrote the first draft of this book shortly after earning my Ph.D. in physics in 1977. Since I wrote the first draft, I have not seen compelling scientific evidence to support or deny the existence of god. I have seen the facts underlying the scientific worldview change substantially in a single generation. Some of the changes, such as reclassifying Pluto from a planet to a dwarf planet, do not have a major impact on the mainstream scientific worldview. However, significant changes have occurred over the past half-century. For example, changes in physics and astronomy include the discovery that the neutrino has mass, the proliferation of particles and their replacement with fields, and the recognition that ordinary matter is apparently a small percentage of the mass and energy in our universe. With science evolving as it has over the centuries, it is wise to view the scientific worldview as a work in progress.

❁ ❁ ❁

Historically we have seen the belief in many gods replaced by the belief in a single god, godhead, or no god at all. Philosophy has not provided compelling answers about nature or the existence of god. The scientific method was developed to help us resolve disputes about our understanding of nature. It relies on reason and observation to acquire knowledge and has helped generate insights into nature that have forced both religion and philosophy to adapt. Despite the success of the scientific method, British philosopher Bertrand Russell (1872–1970) offered the following caution:

> **Q14.14.** So far as the physical sciences are concerned, Hume is *wholly* in the right: such propositions as 'A causes B' are

> never to be accepted and our inclination to accept them is to be explained by the laws of habit and association. [Russell, 1972, p. 669]

We heed Mr. Russell's advice as we apply what we have learned in our quest for truth.

Chapter 15

Competing Worldviews?

Do science and religion offer competing or complementary views of nature? It is difficult to reconcile the differences between the mainstream scientific worldview and some alternative theological explanations if we insist on the literal interpretation of religious texts such as the *Bible*. On the other hand, if we use insights gained from modern science, we find hints of how religion and science can provide complementary views of the universe. We consider different worldviews here by summarizing the scientific worldview before contrasting it with alternative worldviews.[1] We then consider what different worldviews say about the future.

The Scientific Worldview

The scientific worldview is a working theory of the origin and evolution of the physical universe, and the origin and evolution of life. It is natural to begin our description of the origin of the universe at the beginning. Yet, this implies there was a beginning: an implication that has theological implications and is not required by the best observations available today. Was there a beginning? How can we decide if the universe had an origin? We saw in Part 1 that we need to understand how matter behaves on the subatomic scale to understand how the universe began.

A scientific revolution on the subatomic scale occurred when quantum theory replaced classical physics. It set the stage for many more discoveries and helped people realize that answers to our questions about the cosmos must be sought in the subatomic realm as well as in the realm of astronomical observations. Scientists have learned about the subatomic world and attempted to use this knowledge to explain the origin of the universe without reference to the supernatural. Even though the explanations are plausible within the mainstream paradigms of modern science, they are not compelling because they are more speculative and descriptive than predictive.

If we believe that the universe is real, we can ask: How did the universe evolve after the Big Bang? Many cosmological models are considered plausible scientific descriptions of the observable universe. The determination of parameters used in cosmological models is not exact enough to definitively choose the model that applies to our universe. Scientists generally believe that celestial objects such as gaseous nebulae, galaxies, stars, and planets formed from gases permeating the universe after the Big Bang, but the precise mechanisms are unclear.

One of the most important observations we have about life is that all living organisms are based on a common structure: the cell structure. Organisms range in complexity from single all-purpose cells to collections of millions of specialized cells. The simplest living organisms consist of a single cell. These unicellular organisms are believed to be the ancestors of all life.

The origin of the simplest forms of life depends on several factors. Macromolecules need to be formed from smaller molecules by carbon fixing, or the addition of carbon atoms. Once formed, macromolecules must be able to perform several functions, such as serve as a cell-like enclosure with an inside and an outside, convey information, and be self-replicating. A plausible evolutionary sequence for the development of life from inorganic chemicals depends on an increasing level of chemical complexity as we move from inorganic chemicals to the simplest unicellular organisms, but many questions remain. For example, scientists do not know whether the chemistry of the origin of life required improbable reactions working over vast intervals of time, or probable reactions working over a relatively short period of time. Nevertheless, secular scientists assume that the origin of life from inorganic chemicals does not require a supernatural entity, but it does require faith that science will eventually fill in the gaps in our knowledge as we move from one step to another in the evolutionary sequence.

Scientists would like to develop a theory of the evolution of living organisms, from the simplest to the most complex, using only materials present in the physical universe. Much of the scenario of the evolution of life is based on empirical observations. Though the observations are firmly established, their interpretation is not. It is the interpretation of observations and the intellectual leaps across gaps in the observational evidence that are a principal source of controversy between competing worldviews.

Point to Ponder: Plato's Allegory of the Cave

Socrates (ca. 470–399 BCE) presented Plato's (428–347 BCE) allegory of the cave in *Book VII* of The Republic [Cronk *et al.*, 2004, *Book VII*, pp. 118–123; and Plato, 1991]. The allegory of the cave discusses the role of education in the search for truth, and questions the validity and completeness of empirical knowledge.

In the allegory of the cave, prisoners have lived their entire lives chained in a cave. They can only see the cave wall in front of them and cannot see behind them. They do not know about a fire burning behind them or the world outside of the cave.

A barrier stands between the prisoners and the fire. Puppeteers can walk on the barrier and cast puppet shadows on the cave wall in front of the prisoners. The prisoners can see the puppet shadows and are unaware that real puppets are casting the shadows. The prisoners name the shadows and mistakenly talk about the shadows as real objects. Their language confuses shadows with the reality of real objects.

When the prisoners are released, one of the prisoners leaves the cave and sees the real forms of objects in the sunlight. The freed prisoner realizes that the shadows he saw in the cave were not real. He returns to the cave and tries to explain what he saw to the prisoners remaining in the cave. They do not believe him.

One interpretation of the allegory of the cave is that enlightened people (the freed prisoner) have difficulty educating others (the prisoners in the cave). Another interpretation is that the objects we can sense and describe are only shadows and not reality. From this point of view, the prisoners in the cave exemplify people whose empirical knowledge represents only shadows that may obscure a deeper reality.

The Multiverse Worldview

Is god necessary to explain the existence of life and the universe? A scientific consensus is still taking shape, and we can expect some details to change as scientists continue their work. If we take a secular view, we can present a succinct description of a working scientific worldview.

Many mainstream scientists begin with the hypothesis that there is not one universe, but many universes with many different properties.[2] Each of the universes in this vast landscape of universes obeys rational and physical laws. The landscape is a consequence of quantum fluctuations in the vacuum or some other physical mechanism that is, at present, still being sought. Out of this landscape of universes, we just happen to live in one with the properties that allowed intelligent life to emerge.

First, the Big Bang occurred in our universe. It is possible that we are in just one of many universal cycles, but it is sufficient for us to consider only the present cycle. All of the mass and energy in the universe appeared, including observable mass–energy, dark mass, and dark energy. The universe expanded (first inflationary expansion, followed by Hubble expansion). As the universe expanded and cooled, quarks combined to form subatomic particles such as protons and neutrons, which later combined to form atoms and molecules. Slight irregularities in the distribution of matter led to the formation of gaseous nebulae, which became the incubators for stars and galaxies. Matter in the vicinity of stars became the raw material for the formation of planets. Some of the myriads of planets that exist in the universe were suitable for carbon-based life to emerge. Earth was one such planet.

Conditions on the primitive Earth allowed inorganic materials to form organic materials and eventually single-celled organisms. Some single-celled organisms evolved into more complex structures until recently, from a geological perspective, human beings became the dominant species on the planet.

So, is god necessary to explain the existence of life and the universe? If you believe the summary I just gave, god is not necessary to explain the existence of life and the universe.

We must now ask, should we accept the scientific worldview, or have we missed something?

Point to Ponder: Internet Humor — God, the Scientist, and Life

The following story about the origin of life was found circulating on the internet in 2008 (author unknown):

God is sitting in Heaven when a scientist says to Him, "Lord, we don't need you anymore. Science has finally figured out a way to create life out of nothing."

"Oh, is that so? Tell me ..." replies God.

"Well," says the scientist, "we can take dirt and form it into the likeness of you and breathe life into it, thus creating man."

"Well, that's interesting. Show Me."

So the scientist bends down to the earth and starts to mold the soil.

"Oh no, no, no ..." interrupts God. "Get your own dirt."

Part 1 summarized the mainstream attempt by scientists to get their own "dirt."

Genesis and the Origin of the Universe and Life

Scholars have offered differing opinions about biblical creation. In his defense against the inquisition in 1615, Italian astronomer Galileo Galilei (1564–1642) expressed the belief that both science and Genesis contain truth in a "Letter to Madame Christina of Lorraine, Grand Duchess of Tuscany" [Galilei, 1615]. Galileo did not believe that the *Bible* was intended to be a science text. He argued that advancements in science should properly be used to help us interpret Biblical passages that may be ambiguous. On the other hand, theologian Fred Kramer said that "It must be evident to every thinking Christian that Genesis 1 and much of Genesis 2 [the creation accounts] cannot be eyewitness accounts ... Genesis 1 is either, as higher critics assume, the product of human minds, whether you call it a myth, saga, or priestly wisdom; or it is the fruit of divine revelation" [Zimmerman, 1970, p. 106]. More recently, Orthodox Jewish physicist Gerald L. Schroeder (1938–) reinterpreted the original Hebrew texts in the context of modern science and suggested a chronology compatible with the scientific chronology [Schroeder, 1997], while Old Testament scholar John H. Walton (1952–) argued that biblical creation accounts were about creating function, not substance [Walton, 2009]. We consider the interpretations proposed by Schroeder and Walton in more detail to illustrate alternative interpretations.

Schroeder's Interpretation

A literal reading of Genesis 1 says that biblical creation required less than a week. By contrast, the scientific worldview is working with a universal age of about 14 billion years. Can this conflict be reconciled? Can we relate the Genesis timetable to the scientific timetable?

A study of relativity shows that the length of time duration depends on the observer. To a relativistic observer moving at the speed of light, say on the horizon of the expanding universe, 14 billion years of universal evolution would appear to be but an instant. Are earthbound scientists correct in their reckoning of cosmic time, or is the relativistic observer correct? Relativity tells us both are correct from their point of view. From whose point of view was the creation account in Genesis written?

We do not know with certainty whether a day in the first chapter of Genesis was meant to be literally translated, or was intended to be symbolic. In the *Confessions of Saint Augustine*, published around 400 CE, Augustine of Hippo (354–430) raised questions about the meaning of time in Genesis 1 and 2 more than 1,000 years before Charles Darwin (1809–1882) introduced evolution through natural selection. Theologian Raymond F. Surburg emphasized the problem: "One of the key problems in the interpretation of Genesis 1 is the definition of [the Hebrew word] 'yom,' usually translated 'day'" [Surburg, 1959, p. 57]. Surburg cautions us to "… bear in mind that 'yom' occurs no fewer than 1,480 times in the Old Testament and that it is properly translated by over 50 different words, including 'time,' 'life,' 'today,' 'age,' 'forever,' 'continually,' and 'perpetually'" [*Ibid.*, p. 58]. If we assume 'day' was used in a relative sense as physicist Gerald Schroeder suggested [Schroeder, 1997], how do we compare the scientific and Genesis versions of the origin of the universe and life?

Schroeder's chronology is presented in Table 15.1 for comparison. Time is expressed in millions of years before present (MYBP). A comparison of

Table 15.1 Schroeder's view of the origin of the universe and life [after Schroeder, 1997, p. 67].

Day (Genesis 1)	Start of Day (MYBP*)	Description of Main Event(s) of the Day	
		Bible	**Science**
1	15,750	The creation of the universe; light separates from dark (Gen. 1:1–5)	The Big Bang marks the creation of the universe; light literally breaks free as electrons bond to atomic nuclei; galaxies start to form
2	7,750	The heavenly firmament forms (Gen. 1:6–8)	Disk of Milky Way forms; Sun, the main sequence star, forms
3	3,750	Oceans and dry land appear; the first life, plants, appear (Gen. 1:9–13)	The Earth has cooled and liquid water appears 3.8 billion years ago followed almost immediately by the first forms of life: bacteria and photosynthetic algae
4	1,750	Sun, Moon, and stars become visible in the heavens (Gen. 1:14–19)	Earth's atmosphere becomes transparent; photosynthesis produces oxygen-rich atmosphere
5	750	First animal life swarms abundantly in waters; followed by reptiles and winged animals (Gen. 1:20–23)	First multicellular animals; waters swarm with animal life having the basic body plans of all future animals; winged insects appear
6	250	Land animals; mammals; humankind (Gen. 1:24–31)	Massive extinction destroys over 90% of life. Land is repopulated: by hominids and then humans

Note: *MYBP is millions of years before present.

the order of the events shows reasonable agreement between the Genesis events and the scientific events in this interpretation.

Part of the problem with this type of analysis is that we are reinterpreting the meaning of words so that we can match current scientific interpretations. It is possible that the new meanings are correct and Genesis does contain a chronological sequence that is compatible with the scientific chronology of the origin of the physical universe and life. Critics will say, however, that this approach shows that religious teachings are not immutable truths; they are subject to human interpretation.

Walton's Cosmic Temple Inauguration Interpretation

Old Testament scholar John H. Walton's Cosmic Temple Inauguration (CTI) interpretation [Walton, 2009] was introduced in Chapter 12. Schroeder's interpretation of Genesis 1 outlined in Figure 15.1 is based on a reinterpretation of the meaning of the Hebrew word "yom," while Walton's interpretation of Genesis 1 is based on the reinterpretation of the Hebrew word "bara." The CTI interpretation suggests that the 7 days of Genesis 1 "are not given as the period of time over which the material cosmos came into existence, but the period of time devoted to the inauguration of the functions of the cosmic temple" [*Ibid.*, p. 92].

According to Walton, "on day 1 God created the basis for time; day 2, the basis for weather; and day 3, the basis for food. These 3 great functions — time, weather, and food — are the foundation of life" [Walton, 2009, p. 59]. Given the foundation of life, Walton said that in "days 4 to 6 the functionaries of the cosmos are installed in their appropriate positions and given their appropriate roles" [*Ibid.*, p. 71]. Day 7 of Genesis 1 is the most important, says Walton, because day 7 is the day that God takes up his dwelling in the midst of the cosmic temple:

> **Q15.1.** Genesis 1 can now be seen as a creation account focusing on the cosmos as a temple. It describes the creation of the cosmic temple with all of its functions and with God dwelling in its midst. This is what makes day 7 so significant, because without God taking up his dwelling in its midst, the (cosmic) temple does not exist. The most central truth to the creation account is that this world is a place for God's presence. [Walton, 2009, p. 84]

In the CTI interpretation, eternal God was at work before the inauguration of the cosmic temple:

> **Q15.2.** Prior to day 1, God's spirit was active over the nonfunctional cosmos; God was involved but had not yet taken up his residence. The establishment of the functional cosmic temple is effectuated by God taking up his residence on day 7. [Walton, 2009, p. 84]

Beware the God of the Gaps

Parts of the scientific worldview rely on faith to bridge gaps in knowledge and understanding. The two main topics of our discussion — the origin of the universe and the origin of life from inanimate material — are examples of gaps that have yet to be filled by scientific knowledge and understanding. It is tempting for some people to use god to explain things that the science of the day could not explain. This has been referred to as god of the gaps.

Galileo's dispute with the Catholic Church and Darwin's theory of evolution by natural selection are two historical examples of scientists providing new explanations for natural phenomena that challenged the religious teachings of their day. As science matured, rational explanations were found to fill in gaps in knowledge and understanding. Religion and religious teachings adapted. The adaptation, however, could appear to someone searching for truth as a change in beliefs that were supposed to be divinely inspired and infallible. The result was a loss of credibility for religious teachings. People who choose to believe in a god need to be careful that their beliefs are not based on incomplete or transitional knowledge, such as the shadows in Plato's allegory of the cave.

What Does the Future Hold?

We are all mortal. Within a century most of us will be gone. Our Solar System will last much longer, but even its viability is limited.[3] The future evolutionary course of the Sun is depicted by the line in Figure 15.1. This figure is a plot of star luminosity versus the effective temperature of a star. Plots like Figure 15.1 were first made by Danish astronomer and chemist Ejnar Hertzsprung (1873–1967), and American astronomer Henry Norris Russell (1877–1957). They are known as Hertzsprung–Russell diagrams and are useful tools for understanding stellar evolution.

An evolutionary sequence for our Sun has been sketched on the Hertzsprung–Russell diagram (Figure 15.1). The figure shows that our Sun is presently in the Main Sequence of stars. According to this evolutionary sequence, a few billion years from now, perhaps five, our Sun will exhaust its hydrogen. As this nuclear fuel depletes, the solar core will shrink and

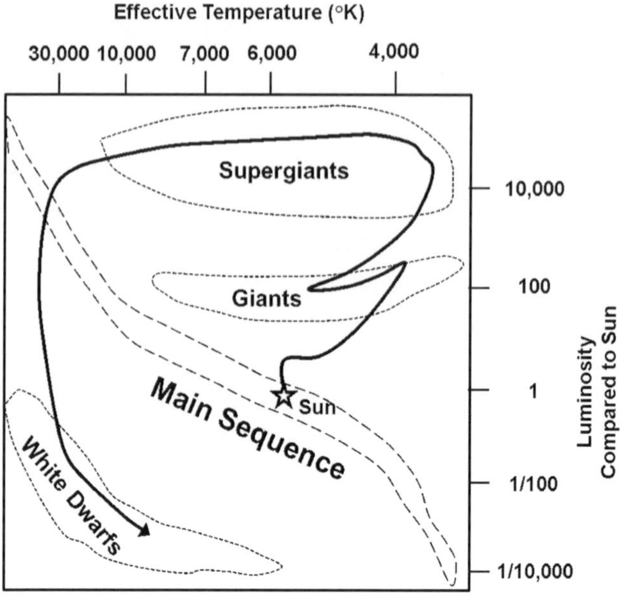

Figure 15.1. Hertzsprung–Russell diagram and the future of the Sun [ATNF, 2021].

the solar surface will expand. Temperatures will rise in the shrinking core; the surface of the Sun will expand and brighten. Eventually, the expanding stellar surface will engulf the inner planets, including Earth, as the Sun becomes a red giant star. A helium flash will initiate the fusing of helium in the Sun and increase the temperature of the Sun along what is known as the "horizontal branch." In time, the Sun will exhaust its remaining hydrogen nuclear fuel as it follows the Asymptotic Giant Branch (AGB) from the set of giant stars to the set of supergiant stars. The Sun will eventually transform into a white dwarf as the radiance of energy from the Sun diminishes. The final stage of the Sun will be a black dwarf, that is, a white dwarf that can no longer emit significant heat or light.

Cosmologically, the future of our universe depends on dark energy. Physicist Michael S. Turner (1949–) succinctly summarized the dependence in 2003:

> **Q15.3.** Dark energy leads to a revision in our view of cosmic density. With matter alone, destiny and geometry are one: Closed universes re-collapse and open or flat universes

> expand forever. If dark energy is vacuum energy, our flat universe will continue accelerating to a bleak future — in 100 billion years, all but a few hundred galaxies nearby will have their light shifted too far into the red to be seen. If dark energy eventually dissipates, the universe will begin to decelerate, possibly even re-collapse. [Turner, 2003, p. 11]

If we find our universe is in fact closed and oscillating, the oscillating universe will reach a point when expansion is no longer possible. From that time on, the universe will contract until all matter–energy is confined to an infinitesimally small point: a new singularity. Then the cycle can begin again.

Our personal mass–energy will participate in whichever universe is real. It appears that doom will be the ultimate fate of our species in the natural universe. Will we either freeze or burn according to the evolutionary laws regulating an impersonal universe, as atheism implies? Or will we someday be freed from our universal confines? What of our immortal self, if it exists?

Religions generally agree that, in the end, the purpose of human life is to acquire knowledge of the Godhead.[4] For example, reincarnation, as espoused by Hinduism, gives us many chances to attain this knowledge. Christianity, by contrast, gives us only one chance but speaks of a forgiving God. If we attain knowledge of the Godhead, we will attain a supernatural haven such as Heaven. Can we use cosmological evidence to decide which religious path should be chosen?

According to cosmology, an open universe implies the universe has not always existed: it began at some event in space–time. The universe, in other words, was created. Western religions tell us the creator was God, or Allah. Eastern religions tend to favor a closed or oscillating universe that does not need a creator. In both cases, we could take a deist view and argue that the laws governing the universe and our existence within it had to be created. The creator's role may not require creating the mass–energy of the universe, but it would require creating the physical laws. A theist would then argue that we, as human beings, should be able to have a personal relationship with this creator.

Alternatively, an atheist could regard our existence as a mere accident: just the realization of one possible quantum state on a universal scale. The entire universe may then be the outgrowth of a single quantum process that took an unlikely path, or it may simply be one of a multitude of universes in a multiverse. If we accept this point of view, we can justify atheism. If we are unwilling to choose from among the competing worldviews given the uncertainties in our knowledge, we could take an agnostic position and say we just do not know.

The cosmological evidence is inconclusive. We could try to use existing cosmological evidence to decide what type of universe we live in, and possibly determine the path to knowledge of the true god, but a decade or a generation from now the cosmological evidence may indicate a different result as technology improves and more data are collected. We would once again have many choices to make with no guarantees that our choices would continue to be valid with the discovery of new information.

❁ ❁ ❁

The modern scientific worldview of the origin of the universe and the evolution of life has great appeal to the scientifically oriented mind. Novelist Dan Brown (1964–) used fiction to explore the credibility and authority of Christianity, particularly the Roman Catholic Church, and to declare that the concept of God is obsolete. In *Angels and Demons*, Brown used a camerlengo, an official of the Papal Court, to state his case:

> **Q15.4.** Science is the new God. Medicine, electronic communications, space travel, genetic manipulation … these are the miracles about which we tell our children. These are the miracles we herald as proof that science will bring us the answers. The ancient stories of immaculate conception, burning bushes, and parting seas are no longer relevant. God has become obsolete. Science has won the battle [with religion]. [Brown, 2000, pp. 378–379]

Through his camerlengo, Brown wrote that "science's victory has cost every one of us… Science proclaims that Planet Earth and its inhabitants are a meaningless speck in the grand scheme. A cosmic accident… [Science] shatters God's world into smaller and smaller pieces in quest of meaning … and all it finds is more questions" [Brown, 2000, p. 379].

Chapter 16

Theism or Atheism?

Our quest has carried us far. Beginning with the Big Bang, we traced the expansion and corresponding cooling of the universe; the formation of the galaxies, stars, and planets, including Earth; the origin of unicellular life from inorganic chemicals; the evolution of multicellular life from unicellular life; and ultimately the appearance of Homo sapiens. Except for initiating the Big Bang, we have not needed to include a supernatural factor (a deity) into our discussion. Indeed, if we accept the notion of creation from quantum foam, the existence of a landscape of universes, or some other natural explanation, then we do not need a creator at all, at least not for the empirical aspects of our universe.

Along the way we have encountered gaps in our knowledge. We have built bridges across the gaps with faith and reason. We have considered many issues, but other issues remain to be addressed as we seek to answer the question: does god exist? In this chapter, we consider three philosophical arguments for the existence of god, and discuss the perspectives of atheists and humanists. We then examine the question of God's existence by examining potential consequences associated with our answer.

Three Arguments for the Existence of God

Three historically significant philosophical arguments for the belief in god are the ontological argument, the cosmological argument, and the teleological argument or argument from design.[1] Each of these arguments is described and then discussed relative to the framework of the scientific worldview.

The Ontological Argument

The ontological argument for the existence of God was first formulated by Anselm (1033–1109), an Archbishop of Canterbury. It begins by defining God as the most perfect conceivable being that depends on no other being. Anselm argued that, since it is possible for us to conceive of a most perfect being that depends on no other, the being must exist. The existence of God is self-evident based on the definition of God. For example, if we define God as the greatest thing that can be conceived, then the existence of God is deduced from the definition of God. Another form of the ontological argument asserts that if we can conceive of the existence of a necessary being, then that being must exist. This argument has been criticized because it assumes that qualities attributed to God imply the existence of god.

The Cosmological Argument

The cosmological argument for the existence of God is usually attributed to Thomas Aquinas (1224/5–1274). According to the cosmological argument, the existence of the universe and everything in it is contingent on a source beyond itself. God is the noncontingent reality that is needed to explain the reality that we experience. Although cosmological arguments rely on the existence of things, they do not require creation in time.

English philosopher Richard Swinburne (1934–) considered the cosmological argument as a C-inductive argument, that is, an argument that increases the probability that a hypothesis is correct. In this case, the hypothesis is that there is a God. Swinburne summarized his discussion of the cosmological argument by saying "There is quite a chance that, if there is a God, he will make something of the finitude and complexity of a universe. It is very unlikely that a universe would exist uncaused, but rather more likely that God would exist uncaused. Hence the argument from the existence of the universe to the existence of God is a good C-inductive argument" [Swinburne, 2004a, p. 152]. English philosopher Antony Flew (1923–2010) analyzed critical and supportive views of the cosmological argument and concluded that Swinburne's cosmological argument "provides a very promising explanation, probably the finally right one" [Flew, 2007, p. 145].

Point to Ponder: Can You Change Your Beliefs?

Philosopher Antony Flew was a leading proponent of atheism for most of his career. In his 2007 book, *There Is a God*, Flew explained how he changed his mind. Flew summarized his theistic beliefs by saying that "I now believe that the universe was brought into existence by an infinite Intelligence. I believe that this universe's intricate laws manifest what scientists have called the Mind of God. I believe that life and reproduction originate in a divine Source" [Flew, 2007, pp. 88–89].

Flew said that his new beliefs reflect the world picture "that has emerged from modern science. Science spotlights three dimensions of nature that point to God. The first is the fact that nature obeys laws. The second is the dimension of life, of intelligently organized and purpose-driven beings, which arose from matter. The third is the very existence of nature. But it is not science alone that has guided me. I have also been helped by a renewed study of the classical philosophical arguments"[2] [Flew, 2007, pp. 88–89].

The Teleological Argument

The teleological argument for the existence of God, which is also known as the design argument, supports the idea that there is an apparent order in nature, and that the observed order would not be possible without guidance. The argument from design is the oldest of the three basic arguments for the existence of God. Elements of the argument can be found in the writings of Anaxagoras (500–428 BCE), and Reese [1999; p. 761] attributed its first statement to Plato (428–347 BCE). The teleological argument says that God must exist to provide the order we see in the universe.

A more modern version of the design argument was introduced by British philosopher William Paley (1743–1805) in 1802 using the watchmaker analogy. If we are strolling in a field and see a rock, we are not surprised because the rock is in its natural setting. On the other hand, if we are strolling in a field and see a watch, we tend to infer that someone had to put it there because the watch is too sophisticated to be a natural part of the field. Similarly, the complexity of life, especially human life, suggests that everything in the world was made so that we can live in the world. If the world was just a little bit different, then we could not live in it. Writer Larry Witham [2003] summarized the argument for design. He said that theists "point to four kinds of evidence for divine design: the simplicity of nature's laws, the beauty of mathematics and the physical order, the intelligibility of the underlying order of things, and the fact that humans are able to apprehend all these things" [Witham, 2003, p. 216].

The design argument tends to appear in more treatments of modern cosmology because it can be associated with testable hypotheses. As an illustration, the anthropic principle found in textbooks[3] on cosmology asserts that the universe must have those properties that allow life to develop within it at some stage in its history. The anthropic principle is a design argument. From a scientific perspective, the anthropic principle constrains the types of models of the universe that will work given the observed values of fundamental physical constants. Scientific models of the universe must be able to account for the fact that we are here.

Phillip E. Johnson (1940–2019) introduced another version of the design argument called Intelligent Design in his book *Darwin on Trial.* Johnson [1991] argued that the conditions needed for life to emerge in nature are so rare and specific that life could not have occurred by a series of chance occurrences. Other authors, such as Ward and Brownlee [2000] in their book *Rare Earth*, and Ross [2016] in his book *Improbable Planet* have analyzed many of the scientific details associated with the emergence of life. Francis S. Collins (1950–) cautioned theists not to rely too heavily on arguments that use god to fill in gaps in scientific knowledge [Collins, 2006]. These "god of the gaps" arguments can fail as science progresses, and their failure can cast doubt on other religious teachings.

Each of the arguments for the existence of god discussed above can be challenged. For example, in *The God Delusion*, biologist and atheist Richard Dawkins (1941–) presented counter arguments but focused most of his attention on the design argument [Dawkins, 2006]. Dawkins said that the appearance of design is illusory. Referring to Paley's watchmaker analogy, Dawkins argued that the process of natural selection could provide the appearance of sophisticated design over a long expanse of time. He said humans are tempted to hypothesize the existence of an intelligent designer, but the temptation is misdirected because we cannot say who designed the designer. The alternative to an intelligent designer in biology is to say the appearance of design was due to natural selection during biological evolution. Dawkins formulated a god hypothesis and an alternative hypothesis. Dawkins' god hypothesis says "there exists a superhuman, supernatural intelligence who deliberately designed and created the universe and everything in it, including us" [*Ibid.*, p. 31]. His alternative hypothesis says that "any creative intelligence, of sufficient complexity to design anything, comes into existence only as the end product of an extended process of gradual evolution" [*Ibid.*, p. 31]. Dawkins argued that it is preferable to believe that science will eventually answer all outstanding questions rather than to believe in the supernatural, which he believes tends to stifle free inquiry. Dawkins' alternative hypothesis is suitable as a central tenet of atheism.

Personal Observation: Is Theism a Hindrance to Science?

Some scientists argue that a belief in god hinders a person's ability to conduct a scientific research program. After all, if you are willing to believe in the superstition of god, you must be willing to believe anything and are incapable of making sound judgments on the evidence? On the other hand, if you are unwilling to have an open mind about the possibility that something may exist that we cannot observe with our senses or the tools that allow us to extend our senses, then perhaps your mind is closed to following the evidence wherever it leads.

The study of nature using the scientific method has historically been a way for some scientists to learn about god's universe [Pearcey and Thaxton, 1994]. Suppose we can extend the frontiers of science until what is now supernatural becomes part of the nature we observe. Perhaps this is what is happening with the search for more universes or more dimensions. Will scientists someday be able to directly observe a physical god? Is this a ridiculous question, or does it display a willingness to follow the evidence without fear or bias?

Atheism

"Atheism," said French revolutionary and theist Maximilien Francois Marie Isidore de Robespierre (1758–1794), "is aristocratic. The idea of a Supreme Being who watches over oppressed innocence and punishes triumphant crime is essentially the idea of the people" [Wells, 1920, Volume 2, p. 335].

Karl Marx (1818–1883) believed that religion was used by the ruling class to keep the working class acquiescent to its desires. He wrote "religion is the opium of the people" in his 1844 paper "Toward the Critique of the Hegelian Philosophy of Right." Views such as Marx's were the basis of the official acceptance of atheism by the Communist Party of the former Soviet Union. With the fall of the Berlin Wall and the dissolution of the Soviet empire at the end of the twentieth century, religion was once again embraced by many residents of the former Soviet Union. Why was atheism rejected so quickly when it lost its support among those in power? Did the Russian people simply reject atheism because it was forced upon them, or did they reject the notion of atheism? What is atheism?

Simply speaking, atheism is the denial of the existence of any deity.[4] German philosopher Ludwig Feuerbach (1804–1872) believed god was but a projection of human beings: "The divine is nothing else than the human being, or, rather, the human nature purified, freed from the limits of the individual man, made objective ..." [Hick, 1970, p. 160]. Besides identifying divine qualities as mere idealizations of human qualities, Feuerbach claimed that god could only exist if god had qualities which were determinate. If such qualities were not allowed, god's existence was denied: "The denial of determinate, positive predicates concerning the divine nature is nothing else than the denial of religion ... it is simply a subtle, disguised atheism" [Hick, 1970, pp. 160–161].

By contrast, Danish philosopher Søren Kierkegaard (1813–1855) suggested that god was whatever could not be known by reason: "... what is this unknown something with which the Reason collides when inspired by its paradoxical passion, with the result of unsettling even man's knowledge of himself? It is the Unknown. It is not a human being, insofar as we know what man is; nor is it any other known thing. So let us call this unknown something: GOD" [Hick, 1970, p. 175]. Kierkegaard argued that god could not be known by rational proof. Rather, belief in god's existence required a "leap" of faith. If we embraced a rationalistic view that everything in nature could be explained by Reason, then Kierkegaard's god — anything that is unknown to Reason — would vanish.

Feuerbach and Kierkegaard are among the philosophers who have argued that the existence of god is not necessary to our understanding of nature. Can we dispense with the concept of god, as atheism requires, and still understand the entirety of nature?

Originally the existence of deities, and eventually of one god, was motivated by humanity's desire to explain the unknown, especially the existence of itself and its world. For the sake of argument, let us suppose that we do not exist. In this case, there is no need for a creator and one of the most powerful motivations for the postulate of the existence of a deity vanishes. Denying our existence, however, is a very difficult idea to sell. Like French philosopher and mathematician René Descartes (1596–1650), we need only say "I think, therefore I am" and most of us are ready to believe we have disproved the postulate "I do not exist."

Rather than adopting the assumption that we do not exist as the basis for atheism, another possibility is to eliminate the need for creation of the universe altogether. In this case, we can argue that the universe has always existed and that there has never been a need to create it. Some ancient creation myths anticipated this possibility by asserting that the universe began in a state of chaos, or disordered matter, and that the universe we know was formed by bringing order out of chaos. Today, one class of modern cosmological theories known as oscillating universe models avoids the need for creation by assuming that the universe has always existed in one form or another. A related class of cosmological theories refers to a set of universes in a many-universe, or multiverse, model. If we use the term cosmos to include all relevant universes, these theories provide a large-scale cosmological description of the cosmos without requiring a creator. By contrast, philosopher Richard Swinburne rejected the multiverse concept and argued that "it is crazy to postulate a trillion (causally unconnected) universes to explain the features of one universe, when postulating one entity (God) will do the job" [Swinburne, 2004b].

Philosophically, the tenets of atheism imply that only phenomena that can be explained rationally can exist or, conversely, that all existing phenomena can be explained rationally. We cannot scientifically account for an irrational event that may occur only once, such as a miracle. Francis S. Collins defined a miracle as "an event that appears inexplicable by the laws of nature and so is held to be supernatural in origin" [Collins, 2006, p. 48].

Events that occur only once and cannot be reproduced, or are considered highly unlikely, are difficult to explain rationally. However, our ancestors would have found it very difficult to explain many things for which we have since found explanations. We might be able to discuss the probability that the event could occur, but we would be imposing a rational mathematical system — probability theory — on our evaluation of the event. Further, we could study the event and attempt to recreate it in the hopes of determining its cause. This raises the question: how can we decide whether an event was rational or irrational? Another way to phrase the question is to ask: is the event scientifically meaningful? The topic of data validity is an ongoing issue in science. It appears that an atheistic view of the cosmos would be supported by a cosmos that has always existed and that contains only events that can be explained rationally, just as religion has been supported by phenomena that have not been readily explained.

Atheist arguments are numerous and disparate, as a casual search of the internet or the literature will show. Arguments against the existence of god, like arguments for the existence of god, have flaws that can be disputed. We illustrate these arguments below.

Empirical arguments against the existence of god use empirical data to try to disprove the existence of god. For example, the argument from poor design argues that there are countless flaws in the "design" of life forms that exist on Earth. These flaws are imperfections that a perfect god would not have allowed. This argument might be effective in arguing against a perfectly created, static universe, but it does not debunk the idea that god simply put the rules of the universe in place and allowed it to develop on its own.

Another empirical argument, the parsimony argument, can be summed up by the phrase "the simplest explanation is often the correct one," or simpler, "less is better." Similar in theory to Ockham's razor, this argument states that our scientific conception of the universe has progressed to the point where we can reasonably explain creation without need for a god. Some scientists believe that the mainstream scientific worldview contains evidence to support this argument. Clearly, there are missing pieces of the scientific worldview, but those who accept the parsimony argument believe we will one day be able to find answers to all of the lingering scientific questions without needing to resort to the use of god. One might also argue that the cosmological argument for the existence of god, which defines god as the uncaused cause, or "first mover," of the universe, can be simplified by making the universe itself the first mover, leaving out the need for any theistic powers attributed to god, such as omnipotence, self-awareness, or will.

The second type of argument against god's existence is the deductive argument. These arguments use deductive reasoning to disprove god's existence. Their weakness is that they largely rely on the definition of god and thus can be invalidated by changing the definition of god. One deductive argument is known as the omnipotence paradox. This argument attempts to disprove god's existence by arguing that omnipotence is impossible. This is often done by asking a paradoxical question about god's abilities, such as "can god create a rock so large even god cannot lift it?" If god can create such a rock, doing so would end god's omnipotence. If god cannot create such a rock, then god was not omnipotent in the first place.

Another vein of the deductive argument involves the problem of god's omniscience versus the existence of free will. How can god know all and yet people still have free will? Why would a caring god allow the existence of evil and of suffering? Furthermore, if god is all-powerful, why are there those who disbelieve the existence of god? The conflict between omniscience and free will is discussed at length in the next chapter. These arguments tend to be more readily accepted by those who have gone through a difficult time or have felt abandoned by god. An alternative view is that evil results from the absence or rejection of god. Nevertheless, a "crisis of faith" can change one's mind about the existence of god.

The third type of argument is the inductive argument, which attempts to disprove god's existence through inductive reasoning. A prime example of this type of argument is the "no reason" argument, which posits that god has no reason to create the universe or to do anything for that matter. For god to have desired the creation of the universe implies an imperfection or weakness in god because to desire something is to lack something. If god is perfect or omnipotent, then god lacks nothing. Since the universe exists, god must not exist.

Point to Ponder: Can You Cover All Bets?

Suppose you do not want to believe in god for some reason. You may think philosophy and science point to a universe without god. Perhaps you have accepted a reason that has nothing to do with philosophy or science. For example, some people may be concerned that expressing a belief in god can hurt their careers or their social life. You may be unwilling to make a commitment, and are really more of an agnostic than an atheist. Or perhaps you follow the tenets of a religion that is the wrong religion. Is there a way to cover all bets? Can you reject god and religion in this life, or choose the wrong religion, and still enjoy a desirable afterlife? Perhaps you could live a life that satisfies the moral and ethical values common to most major religions without participating in aspects of religion that may not appeal to you, such as tithing or making a pilgrimage to Mecca.

If you choose to live this way, you are gambling that if the religious god exists, your transgressions against the true religion will be forgiven because you lived an otherwise exemplary life. In this way, you can live the life you want, up to a point, and still have a shot at a desirable afterlife. Your gamble may pay off. On the other hand, it may not. What should we believe?

Different religions have different answers. For example, in the New Testament of the *Bible* [NIV *Bible*, 2002; see also Lucado, 2007], John 3:16 quoted Jesus prescribing the path to God: "For God so loved the world that he gave his one and only Son, that whoever believes in him shall not perish but have eternal life." What happens if you choose not to believe? Dante Alighieri's (1265–1321) *Inferno* in *The Divine Comedy* included a special place in Hell for people who did not believe in the Christian God [Kline, 2000].

The final type of argument is the subjective argument, which relies on the experiences of witnesses. This may include those who have experienced some event in their lives that led them to doubt the existence of god, but can also include the word of those who do believe in god. The conflicting accounts of those who claim to have been contacted by god or one of god's messengers can be used as evidence against the existence of god and the validity of those experiences. If god exists, why would god tell one person one thing and another person something else? Or have people somehow garbled the message?

Humanism

Secular humanists believe humanity can be moral and achieve self-fulfillment without believing in god.[5] They do not see the need for the supernatural or the immaterial because their worldview embraces only materialism and naturalism. By contrast, atheists reject the belief in god. Sociologist Phil Zuckerman (1969–) presents a supportive view of secular humanism and concludes that secular humanism offers "a heavy dose of reality, tinged with hope based on experience and love" [Zuckerman, 2020].

Humanistic psychologists such as Abraham Maslow (1908–1970) and Carl Rogers (1902–1987) emphasized the uniqueness of the individual and the importance of self-actualization. They believed that human beings have free will and are born with a desire to fulfill their potential. Secular humanism embraces humanistic psychology but divorces itself from the influence of religion.

Corliss Lamont (1902–1995), past President of the American Humanist Association, defined humanism as the "Philosophy of joyous service for the greater good of all humanity in this natural world and advocating the methods of reason, science, and democracy" [Lamont, 1990]. Table 16.1 summarizes the humanistic viewpoint published by Lamont.

Table 16.1 Lamont's humanistic viewpoint [Lamont, 1990].

• People have but one life to lead and should make the most of it in terms of creative work and happiness
• Human happiness is its own justification and requires no sanction or support from supernatural sources
• In any case the supernatural, usually conceived of in the form of heavenly gods or immortal heavens, does not exist
• Humans, using their own intelligence and cooperating liberally with one another, can build an enduring citadel of peace and beauty upon this earth

The humanistic school of thought underlies many modern education programs that include the elevation of a child's self-esteem as one of their most important tasks in the holistic education of a child. The American philosopher and educator John Dewey (1859–1952) is a humanist who rejected the notion of supernatural existence, yet embraced religious values. American educator Anita Woolfolk pointed out that humanistic education stresses the importance of intrinsic motivation and emphasizes personal freedom, choice, self-determination, and the striving for personal growth [Woolfolk, 2019].

Pascal's Wager

French scientist and philosopher Blaise Pascal (1623–1662) made important contributions to fluid mechanics and mathematics. As a child prodigy, Pascal wrote a treatise on geometry and later made contributions to probability theory. In honor of his contributions, the unit of pressure in the International System of Units (which is widely known as the metric system) was named the pascal. But a mystical experience[6] on November 23, 1654 changed Pascal's life and inspired him to provide a rational argument for the existence of the Christian God. Pascal's philosophical ideas were captured in *Pensées* (literally, "thoughts"), a collection of arguments designed to defend his Christian faith. Best known of these arguments is Pascal's Wager.

Pascal observed that we all have to consider the choice: God exists, or God does not exist. Pascal was referring to the Christian God, but the same question applies to other supernatural entities. As a Christian, Pascal assumed that belief in God would be rewarded by eternal life in an afterlife. If this is not the case, then the question of god's existence remains, but the consequences of not believing will change. For example, Islam teaches that we can exercise free will. On the Day of Judgment, each of our souls will be judged and our happiness in the afterlife will depend on how well we adhered to the tenets of Islam.

The choice of whether god exists or does not exist is not a choice that can be decided by debate. Some people — some believers, some nonbelievers — think that they only have to win the debate to validate their worldview. It is possible to win the debate and be wrong about reality. We have a clear choice: either god exists, or god does not exist. Our arguments may change our perception of reality, but they will not change reality. It is a matter of fact whether god exists or not. Pascal pointed out that "a game is being played at the extremity of this infinite distance [between believers and nonbelievers] where heads or tails will turn up. Which will you wager?" [Pascal, 1979, p. 68].

Pascal's cost-benefit analysis places much greater value on the possibility of life after death than the relative importance of the costs or benefits we acquire during our lifetimes. Pascal's wager can be analyzed using the decision matrix in Table 16.2.

We can understand the essence of Pascal's argument by considering the wager from Pascal's point of view. Two of the four possibilities in Table 16.2 have relatively small impact on our lives. If we wager that God exists, and God does not exist, then we have lost relatively little since the cost of our belief includes behaviors such as worship and tithing. If we wager that God does not exist, and God does not exist, then we gain relatively little since the benefit of our belief is a life free of religious obligation. By contrast, the two remaining possibilities have significant consequences. In particular, if we wager that Pascal's Christian God does not exist, and God does exist, we face eternal damnation. If we wager that God exists, and God does exist, we gain eternal salvation.

Pascal argued that our behavior depends on the question of God's existence: "For it is not to be doubted that the duration of this life is but a moment; that the state of death is eternal, whatever may be its nature; and that all our actions and thoughts must take such different directions, according to the state of that eternity" [Pascal, 1979, p. 61].

Pascal considered the three possible states of belief listed in Table 16.3. Suppose we do not like this set of states. We could say that we are not going to gamble and refuse to choose to believe either that God exists or that God does not exist. Pascal was aware of this option and reminded us that "… you

Table 16.2 Decision matrix for Pascal's wager.

Wager	Reality	
	Pascal's God Exists	**Pascal's God Does Not Exist**
God exists	Gain eternal salvation	Small cost
God does not exist	Suffer eternal damnation	Small benefit

Table 16.3 **Pascal's possible states of belief.**

Belief	State
1	God does not exist and we are annihilated at death
2	God exists and we are punished in the afterlife for our unwillingness to believe
3	God exists and we are rewarded in the afterlife for our willingness to believe

must wager. It is not optional. You are embarked [in this life]. Which will you choose then?" [Pascal, 1979, p. 69].

In Pascal's time, some people accepted Belief 1 and called it "shaking off the yoke" [Pascal, 1979, p. 59]. By accepting Belief 1, nonbelievers could reject the demands of religion. We saw in Chapter 1 that historian Edward Larson and co-author Larry Witham asked members of the United States National Academy of Sciences if they believed in god [Larson and Witham, 1998]. Larson and Witham found that approximately 72% of survey respondents did not believe in god, 7% expressed belief in god, and the remaining 21% were undecided (either doubters or agnostics). Biologist Richard Dawkins, a contemporary atheist, reported that a similar survey was conducted by R. Elisabeth Cornwell and Michael Stirrat in Britain [Dawkins, 2006, pp. 101–102]. Cornwell and Stirrat surveyed scientists in the British Royal Society and found results that were similar to the results found by Larson and Witham. Many leading scientists in the United States and Britain accept Belief 1 in Table 16.3 and reject god.

As an example, Nobel Prize-winning physicist Steven Weinberg (1933–) lamented in his description of the origin of the universe that "The more the universe seems comprehensible, the more it seems pointless" [Weinberg, 1977, p. 144]. Pascal anticipated Weinberg's lamentation four centuries earlier when Pascal wondered "What joy can we find in the expectation of nothing but hopeless misery? What reason for boasting that we are in impenetrable darkness?" [Pascal, 1979, p. 58].

British physicist Stephen Hawking (1942–2018) was seeking a theory that would eliminate the need for god as the creator of the universe. Hawking believed that the universe could be completely described by a scientific theory that has no need for god. If such a theory could be completed, Hawking observed that "everything in the universe would be determined by the laws of science and by rolls of the dice within the universe" [Hawking, 2001, p. 85]. Hawking confirmed his commitment to this research effort by saying that this "may sound presumptuous, but it is what I and many other scientists believe" [*Ibid.*, p. 85]. Is the scientific evidence for a universe without god so overwhelming that scientists feel compelled to reject the concept of god? Or is their faith in knowledge acquired using the scientific method so strong that they believe any other way of knowing is unnecessary or unacceptable?

A counterargument to Pascal's Wager, known as the Atheist's Wager, argues that instead of believing in god and spending time in the worship of god, one should put that energy into improving the world and the lives of others, for if god exists and god is truly good, this will be just as effective a means of gaining entry into the afterlife as the belief and worship of god. Further, there is the question of which religion is "right." It is a gamble to follow one religion because selecting the "wrong" religion may be worse than having no religion at all.

Finally, a weakness of Pascal's Wager is the assumption that, if god exists, there must be an afterlife. God's existence does not require the existence of an afterlife. There can be a god with no heaven or hell, and without an afterlife, in which case Pascal's benefits for believing in god have changed.

Personal Anecdote: Fear Mongering?

Is Pascal trying to put the "fear of God" into you with his wager? Should you fear eternal damnation if you choose to reject God? On the other hand, suppose God does exist and you will face eternal damnation if you reject God. Was Pascal fear mongering or simply presenting a plausible scenario? Pascal's faith includes a passage from James 5:19–20 in the New Testament of the *Bible* [NIV *Bible*, 2002]: "My brothers, if one of you should wander from the truth and someone should bring him back, remember this: Whoever turns a sinner from the error of his way will save him from death and cover a multitude of sins." A believer may view Pascal's wager as a well-intentioned warning, while a nonbeliever may view his wager as an attempt to instill fear. Once again, perspective is influenced by worldview and, once again, our choice of worldview may have eternal consequences.

Modern atheists rely on the success of the scientific worldview to justify their philosophy, but questions remain. Who or what initiated the Big Bang? Who or what established the physical laws that appear to be serving us so well in our search for understanding? Other questions arise if god does not exist. For example, what is the significance of morality? Are right and wrong arbitrary concepts? Is the universe accidental and meaningless? We now use what we have learned to consider these questions.

Chapter 17

What Is Your Wager?

We are nearing the end of our quest. Before us is a continuum of beliefs to choose from. Table 1.1 and Figure 1.1 illustrate our choices. At one extreme, we can choose to believe that a supernatural being created the universe and is involved in human affairs. At the other extreme, we can choose to believe that there is no god. I asked in Chapter 1 if you could identify where your beliefs would fit in the continuum. Where are you now? Are you ready to make a wager?

Theism and Free Will

One of the major conceptual difficulties of religions such as Judaism and Islam is their belief in both an omnipotent deity and human beings capable of exercising free will. For example, in the *Koran*, the holy book of Islam, we find "Whoever gets to himself a sin, gets it solely on his own responsibility (iv:111); Whoever goes astray, he himself bears the whole responsibility of wandering (x:103)" [Smith, 1965, p. 239]. Eastern religions also embrace free will, as Huston C. Smith (1919–2016) relates: "Karma decrees that every decision must have its determinate consequences, but the decisions themselves are, in the last analysis, freely arrived at" [*Ibid.*, p. 77].

How can we reconcile the apparent incompatibility of an omnipotent deity and human free will? One way is to look at human beings as functioning, psychologically complex organisms. Is our behavior completely predictable in a deterministic sense, or is it indeterminate?

We possess certain physiological traits that appear to function as deterministic systems. These systems include muscular and circulatory systems. Other systems, such as the nervous system (including the brain), are systems whose functions may depend on quantum phenomena.

Quantum theory shows us that the laws of nature are nondeterministic in the sense that we must refer to the state and evolution of an individual system using probabilistic statements. If our behavior does depend on the nondeterministic rules of quantum theory, we can explain the coexistence of human free will with the deterministic powers of an omnipotent deity. In this case, the probabilistic laws of nature are imposed by an omnipotent deity. Since the laws are probabilistic, the behavior of an individual is unspecified. Individually, we can exercise free will. This view does not preclude an omnipotent deity from interfering with the universe it has created. Such interference could appear as a manipulation of the historical probabilities. To us, the manipulation might look like a "miracle." How would we know the manipulation was a miracle and not just the realization of an improbable event like winning the lottery? Considering the present state of our knowledge, we would not know. Identifying an event as a miracle may be difficult to disprove, but is virtually impossible to prove! Thus, from a scientific viewpoint, to claim an event is a miracle is a meaningless claim.

An omnipotent deity can create rules that govern the behavior of large groups of people rather than individuals. From this perspective, we are each free to choose how we will behave. In exercising this freedom, we may cause harm. Within this perspective is a solution to the question of suffering: why does a good and omnipotent god allow suffering?

Adopting the hypothesis that a good and omnipotent god has created rules that constrain groups rather than individuals, individuals are free to make choices that can harm themselves or cause harm to others. Thus, a good and omnipotent god can exercise omnipotence by establishing rules, yet allow freedom of choice by applying the rules to groups, rather than individuals. An analogy from the classroom will help illustrate this perspective. As a teacher, I can establish rules for my students, but each student can choose to ignore those rules or make choices within the rules that still lead to harm. I can intervene at any time, but I may choose not to, even when it is apparent that some harm may occur. Sometimes individual suffering can lead to valuable lessons that will benefit the group. If we adopt this perspective, we no longer have a conflict between free will and a good and omnipotent deity. Instead, we have identified a perspective that allows their co-existence.

Point to Ponder: Evil and the Free Will Defense

Antony Flew (1923–2010) pointed out the "incongruity of evil in a universe created by an omnipotent, all-good Being." Theists responded with "the claim that God gives humans free will, and that all or most of the obvious and scandalous evils are immediately or ultimately due to misuse of this dangerous gift, but that the end results will be the realization of a sum of greater goods than would otherwise be possible." Flew was the first to label this the free-will defense [Flew, 2007, pp. 58–59].

Moral and Ethical Behavior

The existence of god is the basis upon which many people build their moral and ethical systems. If god does not exist, or if we choose to believe that god does not exist, we no longer have an authoritative moral or ethical base. We can replace religious authority with social authority, but the need to consider eternal reward or punishment will no longer be relevant.

People have understood the importance of religious authority for centuries. English lawyer Thomas More (1478–1535), who wrote *Utopia* in 1516, was an advocate for civil tolerance of all religions but not for atheism. More used his fictional King Utopus to explain that atheists should not be entrusted with civil authority because "you may be sure that he [the atheist] will study either with craft privily to mock, or else violently to break, the common laws of his country, in whom remaineth no further fear than of the laws nor no further hope than of the body" [More, 1516, p. 141]. More's King Utopus took a rather dim view of atheism and atheists. The only constraints on the atheist's behavior are personal and societal, which More did not consider to be especially difficult to overcome, particularly when there can be no reward for good behavior in a nonexistent afterlife.

Atheist Richard Dawkins (1941–) argued that mature human beings do not need religious authority to live moral, ethical, and meaningful lives [Dawkins, 2006]. Nonbelievers like Dawkins may be correct in saying that people can live moral, ethical, and meaningful lives without religious authority or rules imposed by god. If you do not believe in god, reincarnation, or an afterlife, you must believe that this life is all that a person can ever experience. From that point of view, you must seek fulfillment in this life because there is no other. Punishment for immorality harms the quality of your life, so an atheist would seek to avoid punishment. Spending your life in jail can be just as dissuasive as spending your life in hell. In either case, the punishment lasts for the duration of the punished person's existence. There may even be a genetic constraint on the social behavior of human beings, but there are enough contemporary examples of heinous crimes and historical examples of violent social movements that make you wonder how much influence such a constraint has on each of us. Similarly, heinous crimes have been committed in the name of religion, demonstrating that even believers are capable of breaking society's rules despite, or in some cases because of, their belief in god.

Point to Ponder: C.S. Lewis and the Meaning of Evil

British writer C.S. Lewis' (1898–1963) experience in the trenches of World War I and his service as an air raid warden during the bombing of Britain in 1940 led to an invitation to give talks to members of the Royal Air Force and, later, deliver a series of BBC broadcasts from 1942 to 1944. His broadcasts were delivered during World War II to a people at war. He chose to speak about his Christian faith and the problems of suffering, pain, and evil. His book, *Mere Christianity*, was prepared from the broadcasts. In a Preface to his book, Kathleen Norris said that Lewis spoke "with no authority but that of experience, as a layman and former atheist" [Lewis, 1952, p. xviii]. Lewis "told his radio audience that he had been selected for the job of describing Christianity to a new generation precisely because he was not a specialist but 'an amateur … and a beginner, not an old hand'" [*Ibid.*, p. xviii].

Lewis recognized that the universe "contains much that is obviously bad and apparently meaningless, but containing creatures like ourselves who know that it is bad and meaningless. There are only two views that face all the facts. One is the Christian view that this is a good world that has gone wrong, but still retains the memory of what it ought to have been. The other is the view called Dualism.

Dualism means the belief that there are two equal and independent powers at the back of everything, one of them good and the other bad, and that this universe is the battlefield in which they fight out an endless war."

Lewis said that "God created things which had free will. That means creatures which can go either wrong or right ... free will is what has made evil possible. Why, then, did God give them free will? Because free will, though it makes evil possible, is also the only thing that makes possible any love or goodness or joy worth having ..." [Lewis, 1952, p. 47].

The Ultimate Gamble

We have sought an answer to the question of god's existence by considering events in the natural world that are the subject of both science and religion. We have used science to provide a plausible worldview of the origin of the universe, the origin of our galaxy, the origin of our Solar System including Earth, the origin of life, and the evolution of human beings. The secular scientific worldview avoids the introduction of a supernatural entity but it contains gaps which require leaps of faith. If you choose to believe the secular scientific worldview, you are putting your faith in the scientific method and in people who adopt and apply the scientific method. You are also choosing to accept many assumptions. The few listed in Table 17.1 are discussed here.

Table 17.1 Selected assumptions of the scientific worldview.

The Universe Is	
A	Rational
B	Understandable
C	Predictable
D	Accidental
E	Purposeless

A. If you accept the scientific worldview, you accept the assumption that the universe is rational. For example, we have found that mathematics can be used to describe natural phenomena. It has been so successful that mathematics has become the language of science. If there are events in nature that are not rational, rationalists tend to dismiss them for a variety of reasons. These events could be temporarily inexplicable, and we assume that we could eventually explain them if we had enough information or a better understanding of nature. Events that seem to be irrational, or actually are irrational, may simply be ignored because they are not reproducible, testable, or falsifiable. Spiritual experiences are usually considered unscientific, irrational, or delusional.
B. If you accept the scientific worldview, you accept the assumption that the universe is understandable. All of the gaps in our scientific understanding of nature will eventually be filled. For example, if the universe had a beginning, we will find a natural explanation for the origin of the universe, such as the formation of matter from quantum foam. If the universe did not have a beginning, then we will learn what matter is and why it has always existed. Dark matter and dark energy will eventually succumb to scientific scrutiny. The steps leading from inorganic material to life will be determined and understood.
C. If you accept the scientific worldview, you accept the assumption that the laws of science are predictable. Scientific laws that have been inferred from past observations will continue to be true in the future.
D. If you accept the scientific worldview, you accept the assumption that the universe is accidental. Chance and time have combined to form the universe as we know it. All of the events that were needed to form life occurred in the right sequence and at the right time. We are a consequence of these events.
E. If you accept the scientific worldview, you accept the assumption that the universe is intrinsically purposeless. Life is an accident without a meaning provided by our materialistic universe. Life's purpose would have to come from within ourselves. There may or may not be other universes. If other universes do exist, their existence is no more or less meaningful than ours.

Could god be responsible for imposing all of the assumptions in Table 17.1? This would actually establish the role god plays in the universe. It is possible that Assumptions A (universe is rational), B (universe is understandable), and C (universe is predictable) are true whether or not god exists. For example, if god exists, the universe may be rational as in Assumption A, or it may include irrational events. Modern religions describe events that science would view as irrational. Some famous examples include Moses (circa 1400 BCE) speaking to God, Muhammad (570–632) speaking to the angel Gabriel, Jesus (ca. 2 BCE–33 CE) rising from the dead and appearing to His disciples, and Pascal's (1898–1963) vision that changed his life. All of these spiritual events have been documented, yet they have no credibility from a scientific perspective because they cannot be reproduced, tested, or falsified. Did these events actually happen, or were they delusions?

It is possible that god exists and assumptions D (universe is accidental) and E (universe is purposeless) are also true, but modern religions typically attribute meaning and purpose in the universe to their conceptions of god. Indeed, an entity that provides meaning and purpose to the universe may actually be a defining characteristic of god.

It is important to realize that the question of god's existence is not the same as asking what god is like. Major theistic religions provide different answers to the latter question given the existence of a supernatural entity. The first question (does god exist?) concerns a philosophical understanding of god, while the second question (what is god like?) asks a religious question that is concerned with the relationship a human being may have with god, if any.

What do you believe about the existence of god? To help you answer this question, consider two simpler questions. First, do you believe that god is possible? If you do not believe that god is possible at this point in our discussion, then you do not believe that god exists. If you think god is possible, then ask yourself the second question: is god needed to explain the universe and our role in it? If you believe that god is necessary, then you believe that god exists. What if you believe that god is possible but not necessary? Agnostics are typically in this position. People who believe in god obviously believe that god is possible even though some may believe that the universe could be explained without god.

We could elaborate on virtually all of the topics discussed so far, but we have enough information now to understand the key issues. At some point, we must confront the question: does god exist? Many people answer the question of the existence of god based on their belief that they have had a spiritual encounter with god. A common example of a direct spiritual encounter is a vision. Another way to decide if you have experienced a spiritual event is to look for spiritual markers in your life. A spiritual marker is a decision point or event that had a significant impact on the course of your life or someone else's life. Spiritual markers are often associated with life-changing experiences, such as graduations, first job, career choices, marriage, the birth of children, divorce, serious illnesses to you or your loved ones, the death of loved ones, and so on. Examples of documented events that some have interpreted as spiritual markers are Blaise Pascal's mystical experience in the seventeenth century [Connor, 2006] and Francis S. Collins' (1950–) treatment of a young farmer in a Nigerian hospital in the modern era [Collins, 2006, pp. 213–218].

The modern scientific worldview is designed to reject subjective observations such as spiritual events and to be skeptical of supernatural interpretations of events like spiritual markers. It is not unusual for science to ignore or reject data points that do not seem to fit into a preconceived structure or cannot be subjected to scientific scrutiny. Indeed, some agnostics and atheists may have experienced what believers would consider a spiritual event and do not recognize it as such because that interpretation does not fit into their worldview. Nonbelievers may interpret such spiritual markers merely as life events, important events, certainly, but simply events in any life that have no deeper meaning. Which is the correct interpretation? That is for each of us to decide.

❂ ❂ ❂

My goal has been to present as objective an appraisal as possible of the information you need to confront the question of god's existence. As we approach the end of our discussion, it is worth repeating an important point: the existence of god does not depend on whether or not we can win a debate. Either god exists, or there is no god.

Each of us should consider the evidence from science and the evidence from our life experience. Some people find it easier to believe in speculative science, such as dimensions that have never been detected or a multitude of universes that have never been observed, rather than in a supernatural being other people claim to have seen. Do you believe that the origin and evolution of the universe and life occurred without divine intervention, or has science shown that divine intervention was needed to achieve such an improbable outcome? Is the universe intrinsically meaningless and pointless, as physicist Stephen Weinberg (1933–) lamented, or does god provide meaning and purpose to the universe?

Amir Aczel (1950–2015) observed that "In so many ways, the same impulse to know the world and our place in it is at the roots of both science and spirituality" [Aczel, 2014, p. 253]. Science writer Nancy Pearcey and chemist Charles Thaxton argued that religion, notably Christianity in Medieval Europe, was "the womb that gave birth to the scientific enterprise" [Pearcey and Thaxton, 1994, p. 19]. Religion professor Paul Brockelman said that scientific cosmology helps us understand the evolution of the universe and gives us a "spiritual vision of a wider order of being to which we belong."[1]

Recognizing that scientists have no definitive knowledge in matters of faith, astronomer Trinh Thuan succinctly summarized our choices: "We can invoke a Supreme Being, who is the author of these [scientific] laws, and who has planned and fine-tuned everything, or else we can come down firmly on the side of chance …" [Thuan, 1995, p. 248]. In his attempt to argue "that science has not disproved the existence of God" [Aczel, 2014, p. 252], Aczel pointed out "that the pursuit of truth should not be driven by zealous agenda" [*Ibid.*, p. 253].

In my quest for truth, I have considered many possibilities and my beliefs have evolved. I know that science has answered many questions, and I expect science to answer even more in the future, but there are gaps in our knowledge, and there are some issues that seem to transcend science. We have learned that at least part of the universe appears to be rational, understandable, and predictable, but is there more? Is the universe merely an accident and our existence meaningless? I am skeptical that sentient life emerged as an accident and without purpose.

When I consider the issues and weigh the evidence, my study of nature and my life experiences have led me to believe that there is a god. Now that we are at the end of our discussion, have you reconsidered your worldview? What have you chosen to believe? Whether we like it or not, each of us must make the ultimate gamble.

Dates of Selected Characters

Name (Last, First)	Dates	Country of Birth
Abraham aka Avraham, Abram or Avram from Ur	ca. 1813–1638 BCE	Babylonia
Aczel, Amir	1950–2015	Israel
Alighieri, Dante	1265–1321	Italy
Anaxagoras	ca. 500–428 BCE	Clazomenae, now Turkey
Anaximander	ca. 610–546 BCE	Miletus, Iona now Turkey
Anselm of Canterbury	1033–1109	Italy
Apollonius of Perga	262–190 BCE	Turkey
Aquinas, Thomas	1225–1274	Italy
Aristotle	ca. 384–322 BCE	Greece
Armstrong, Karen	1944–	England
Arrhenius, Svante	1859–1927	Sweden
Augustine of Hippo aka Saint Augustine	354–430	Algeria
Barr, Stephen M.	1953–	USA
Becquerel, Antoine Henri	1852–1908	France
Berkeley, George	1685–1753	England
Bohr, Niels	1885–1962	Denmark
Bondi, Hermann	1919–2005	Austria
Born, Max	1882–1970	Germany

(*Continued*)

Name (Last, First)	Dates	Country of Birth
de Broglie, Victor	1785–1870	France
Brown, Dan G.	1964–	USA
Brownlee, Donald E.	1943–	USA
Buffon, Georges-Louis Leclerc de	1707–1788	France
Cairns-Smith, A. Graham	1931–2016	Scotland
Capra, Fritjof	1939–	Austria
Carnap, Rudolf	1891–1970	Germany
de Chardin, Pierre Teilhard	1881–1955	France
Chuang Tzu	ca. 369–286 BCE	China
Cloud, Preston	1912–1991	USA
Collins, Francis S.	1950–	USA
Crick, Francis	1916–2004	Britain
Cuvier, Georges	1769–1832	France
Darwin, Charles R.	1809–1882	England
Darwin, Erasmus	1731–1802	England
Dawkins, Richard	1941–	Britain
Democritus of Abdera	ca. 460–370 BCE	Greece
Descartes, René	1596–1650	France
Dewey, John	1859–1952	USA
Doppler, Christian Andreas	1803–1853	Austria
Drake, Frank	1930–	USA
Durkheim, Emile	1858–1917	France
Eddington, Arthur	1882–1944	England
Einstein, Albert	1879–1955	Germany
Eldredge, Niles	1943–	USA
Everett, Hugh III	1930–1982	USA
Faraday, Michael	1791–1867	England
Fermi, Enrico	1901–1954	Italy
Feynman, Richard P.	1918–1988	USA
Fitzroy, Robert	1805–1865	England

Name (Last, First)	Dates	Country of Birth
Flew, Antony	1923–2010	England
Fox, Sydney W.	1912–1998	USA
Frazier, James G.	1854–1941	Scotland
Freud, Sigmund	1856–1939	Austria
Friedmann, Alexander	1888–1925	Russia
Galilei, Galileo	1564–1642	Italy
Gamow, George	1904–1968	Odessa, Russian Empire, now Ukraine
Gautama, Siddhartha aka Buddha	ca. 560 BCE	Nepal
Gell-Mann, Murray	1929–2019	USA
Georgi, Howard	1947–	USA
Gerlach, Walther	1889–1979	Germany
Glashow, Sheldon	1932–	USA
Gold, Thomas	1920–2004	Austrian
Goudsmit, Samuel A.	1902–1978	Netherlands
Gould, Stephen Jay	1941–2002	USA
Haldane, John B.S.	1892–1964	England
Hawking, Stephen	1942–2018	England
Hazen, Robert M.	1948–	USA
Heisenberg, Werner	1901–1976	Germany
Heraclitus	535–475 BCE	Ephesus, now Turkey
Hertzsprung, Ejner	1873–1967	Denmark
Higgs, Peter	1929–	England
Hill, W.E. (William Ely)	1887–1962	USA
Hipparchus of Nicaea	190–120 BCE	Turkey
Hobbes, Thomas	1588–1679	England
Hooke, Robert	1635–1703	England
Hooker, Joseph D.	1817–1911	England
Horowitz, Norman	1915–2005	USA

(*Continued*)

Name (Last, First)	Dates	Country of Birth
Hoyle, Fred	1915–2001	England
Hubble, Edwin P.	1889–1953	USA
Hume, David	1711–1776	Scotland
Hutton, James	1726–1797	Scotland
Jesus of Nazareth	ca. 2 BCE–33 CE	Bethlehem
Johnson, Phillip E.	1940–2019	USA
Josephus, Flavius	ca. 37–100 CE	Jerusalem
Joyce, James	1882–1941	Ireland
Kant, Immanuel	1724–1804	Germany
Kepler, Johannes	1571–1630	Germany
Kierkegaard, Søren	1813–1855	Denmark
Knoll, Andrew H.	1951–	USA
Kuhn, Thomas	1922–1996	USA
Lakatos, Imre	1922–1974	Hungary
Lamarck, Jean-Baptiste	1744–1829	France
Lamont, Corliss	1902–1995	USA
Laplace, Pierre	1749–1827	France
Lao Tzu aka Laozi	ca. 604–531 BCE	China
Leavitt, Henrietta Swan	1868–1921	USA
Leeuwenhoek, Antonie van	1632–1723	Netherlands
Leibniz, Gottfried Wilhelm von	1646–1716	Germany
Lemaître, Georges	1894–1966	Belgium
Leucippus	ca. fifth century BCE	Greece
Lewis, C.S.	1898–1963	Britain
Linnaeus, Carl	1707–1778	Sweden
Locke, John	1632–1704	England
Lorentz, Henrik Anton	1853–1928	Netherlands
Lucado, Max	1955–	USA
Luther, Martin	1483–1546	German

Name (Last, First)	Dates	Country of Birth
Lyell, Charles	1797–1875	Scotland
Malthus, Thomas R.	1766–1834	England
Marett, Robert R.	1866–1943	England
Marx, Karl	1818–1883	Germany
Maslow, Abraham	1908–1970	USA
Mayr, Ernst	1904–2005	Germany
Maxwell, James Clerk	1831–1879	Scotland
Mendel, Gregor Johann	1822–1884	Germany
Mendeleev, Dmitri	1834–1907	Russia
Mill, John Stuart	1806–1873	England
Miller, Stanley L.	1930–2007	USA
Milton, John	1608–1674	England
More, Thomas	1478–1535	England
Moses (biblical)	ca. 1400 BCE	Ancient Egypt
Muhammad, Abū al-Qāsim	570–632	Mecca, Arabia
Nawood, Nessim Joseph	1927–2014	Iraq
Newton, Isaac	1643–1727	England
Ockham, William of	ca. 1287–1347	England
Orgel, Leslie E.	1927–2007	England
Oparin, Alexander I.	1894–1980	Russia
Paley, William	1743–1805	England
Parmenides	ca. 515 BCE	Elea, Greek colony in Italy
Parrinder, Geoffrey	1910–2005	Britain
Pascal, Blaise	1623–1662	France
Pasteur, Louis	1822–1895	France
Penzias, Arno A.	1933–	USA
Planck, Max	1858–1947	Germany
Plato from Athens	428–347 BCE	Greece
Ponnamperuma, Cyril	1923–1994	Sri Lanka
Popper, Karl	1902–1994	Austria

(*Continued*)

Name (Last, First)	Dates	Country of Birth
Powell, Cecil F.	1903–1969	UK
Protagoras	ca. 490–420 BCE	Greece
Redi, Francesco	1626–1697	Italy
Robespierre, Maximilien	1758–1794	France
Rogers, Carl	1902–1987	USA
Rubin, Vera	1928–2016	USA
de Rújula, Álvaro	1944–	Spain
Russell, Bertrand	1872–1970	UK
Russell, Henry Norris	1877–1957	USA
Rutherford, Ernest	1871–1937	New Zealand
Sagan, Carl	1934–1996	USA
Salam, Abdus	1926–1996	Pakistan
Schleiden, Matthias, J.	1804–1881	Germany
Schmidt, Wilhelm	1868–1954	Austria
Schroeder, Gerald L.	1938–	USA
Schwann, Theodor	1810–1882	Germany
Slipher, Vesto	1875–1969	USA
Socrates from Athens	ca. 470–399 BCE	Greece
Smith, Huston C.	1919–2016	China
Smith, William 'Strata'	1769–1839	England
Spencer, Herbert	1820–1908	England
Spinoza, Baruch	1632–1677	Portugal
Stern, Otto	1888–1969	Germany
Swinburne, Richard	1934–	England
Tacitus, Publius Cornelius	ca. 56–120	Gaul
Thomson, William aka Lord Kelvin	1824–1907	Ireland
Tsiolkovski, Konstantin	1857–1935	Russia
Turner, Michael S.	1949–	USA
Tylor, Edward B.	1832–1917	England
Uhlenbeck, George E.	1900–1988	Netherlands

Name (Last, First)	Dates	Country of Birth
Urey, Harold C.	1893–1981	USA
Ussher, James	1581–1656	Ireland
Voltaire aka François-Marie Arouet	1694–1778	France
Wächtershäuser, Günter	1938–	Germany
Wallace, Alfred Russel	1823–1913	Britain
Walton, John H.	1952–	USA
Ward, Peter D.	1949–	USA
Watson, James	1928–	USA
Wegener, Alfred	1880–1930	Germany
Weinberg, Steven	1933–	USA
Wells, Herbert George	1866–1946	England
Wheeler, John Archibald	1911–2008	USA
Whittaker, Robert H.	1920–1980	USA
Wickramasinghe, Chandra	1939–	Sri Lanka
Wilson, Robert W.	1936–2013	USA
Woese, Carl R.	1928–2012	USA
Wöhler, Friedrich	1800–1882	Germany
Ximenez, Francisco	1666–1729	Spain
Yukawa, Hideki	1907–1981	Japan
Zeno	495–430 BCE	Elea, Greek colony in Italy
Zuckerman, Phil	1969–	USA
Zweig, George	1937–	Russia
Zwicky, Fritz	1898–1974	Bulgaria

Discussion Questions

Chapter 1. Quest for Truth

1. Faith: What role has faith played in your life?
2. Have you decided if god exists? If so, what do you hope to gain from reading this book?
3. God's Existence: How would you argue that god exists? How would you refute that argument?
4. Normal Science: List some reasons why mainstream scientists might resist adopting a significant change in normal science.

Chapter 2. Evidence for an Expanding Universe

1. Observations and New Technology: Many advances in science occur when new technology allows us to observe new phenomena. Particle accelerators and telescopes in space are two examples of technology that have expanded our frontiers. Do you think we will ever come to an end of new phenomena? Will the society with the most resources and a willingness to use technology to study nature have a competitive edge in the future?
2. Looking into the Past: Why do astronomers say we are looking into the past when we look at light from distant stars? Do we have any idea what is happening now at those distant stars?
3. Scientific Skepticism: Nineteenth-century physicists once believed in ether. The ether was abandoned after the results of an experiment was unable to detect the ether. Today some scientists believe that other universes exist beside our own. How much faith should we place in theories that rely on unobserved phenomena?

4. Scientific Misconduct: Do you believe plagiarism is misconduct? How should a scientist be treated if the scientist takes someone else's work and does not give credit to the original authors?

Chapter 3. Evidence for Inflationary Cosmology

1. Science and the Supernatural: Some scientists believe that the universe contains more dimensions than the four space-time dimensions of special relativity. Is it possible that the supernatural exists in these other dimensions? If we could discover a way to observe more dimensions, would the supernatural then become part of nature? Is scientific speculation about extra dimensions more valid than belief in a supernatural.
2. Theology and Cosmology: Scientific cosmology speaks of open, flat, and closed universe models. Is it appropriate to test religious views of the cosmos by comparing them to scientific cosmological models? Explain your answer.
3. Speed of Light: Mainstream physics teaches that no massive object can go faster than the speed of light. A simple estimate of the size of the universe is obtained by calculating how far a photon would travel during the life of the universe since the Big Bang. What if it was possible for a massive object to go faster than the speed of light? How would an FTL (faster than light) object affect the size of the universe? What would it do to the estimate of the cosmic energy density, which is obtained by dividing the energy in the universe by the size of the universe?
4. Einstein's Greatest Blunder: Einstein claimed that the introduction of a cosmological constant into his general relativistic equations was his greatest blunder. It now appears that the cosmological constant is needed to understand the behavior of the universe. Under what circumstances should a scientist acknowledge a mistake? Is the failure to acknowledge a mistake unethical? Who has the authority to establish a system of ethics?

Chapter 4. The Big Bang

1. Probability and Uncertainty: If we could know something would happen with certainty, its probability of occurrence would be 100%. When we introduce uncertainty, we have to be concerned about probabilities that are less than 100%. When we gamble, we are making decisions based on the probability that some particular outcome will occur. What are some of the uncertainties we must face in our day-to-day lives? What are some of the risks?
2. Science and Labels: Scientists have inferred that dark matter exists but they do not know what it is. Does applying a label to something make it more credible?
3. Dark Matter and Dark Energy: It now appears that the vast majority of the mass-energy in the universe is not well understood. What effect, if any, does this lack of understanding have on the credibility of scientists who proclaim that we do not need god to understand nature?
4. God and the Big Bang: Do we need God to explain the beginning of the universe? If the Big Bang was simply the beginning of another cycle in a closed universe, would this favor a theistic or an atheistic view of the universe?

Chapter 5. The Inflation Model of Cosmology

1. Time and Space: Did space and time exist before the Big Bang? Does it make sense to talk about the universe before the Big Bang? What is beyond the edge of a finite universe? Is time eternal or finite?
2. Theory of Everything: Do you think science will ever find a theory of everything? Explain your answer.
3. Interpreting a Theory: The interpretation of quantum theories is a disputed topic in physics. Should we believe the results of a theory that can be applied to some experiments successfully but the interpretation of the theory is subject to dispute?

4. Science and Society: Should society support scientists whose work is more theological than scientific? What if their theology is a traditional religion? What if their theology is atheism?

Chapter 6. What Is Life?

1. Life: Define life.
2. Radioactive Dating: Radioactivity was discovered in the late 19th-century. That means we have only known about radioactivity for less than two centuries, yet we use radioactive dating to determine the age of objects that are ancient. Are we assuming that the law of radioactive decay was the same in the past as it is today? What if it was not? How would we know?
3. Atmosphere and Life: Mainstream science believes that the atmosphere and life have affected each other since the earth was young. Do you believe that life affects the atmosphere? Do we have any modern examples? Does the atmosphere affect life?
4. Creation Theory: Is the Big Bang theory a theory of creation? Would you call someone who believes that the Big Bang theory is correct a creationist?

Chapter 7. The Origin of Life

1. Spontaneous Generation: Do you think life began from inanimate matter? Did the origin of life require help from a supernatural being, or could life have been spontaneously generated? If life did occur by spontaneous generation, what is the likelihood that life exists elsewhere in the universe?
2. Biogenesis: Details of the origin of life are still unknown. There appear to be gaps in our knowledge about how experiments like the Miller–Urey experiment are linked to natural environments such as an extremophilic environment. Should we believe that science will fill these gaps?
3. Astronomical Conditions for Life: The conditions on Earth depend in part on the type of Sun we have and our distance from the Sun.

What would happen to life on Earth if the Earth's orbit changed and we moved much closer to the Sun, perhaps as close as Venus? What would happen to life on Earth if we moved close to the orbit of Mars?

4. Panspermia: Some people believe that life on Earth came from outer space. What are some advantages and disadvantages of this belief? How has the study of life under extremophilic conditions altered our view of life on other worlds?

Chapter 8. The Evolution of Life

1. Evolution: Which aspects of evolution are factual? Which aspects are theoretical? Should we say that evolution is either a fact or a theory, or should we say that evolution is both a fact and a theory?
2. Dinosaurs and Fossils: Gerald Schroeder [1997, p. 193] translated Genesis 1:21 as "And God created the big reptiles ..." Schroeder interpreted this translation to mean that Genesis contains a reference to the dinosaurs, the biggest reptiles. Many scientists now believe that the dinosaurs were destroyed by an extinction event. Two extinction events are the Permian extinction and the Cretaceous extinction. Was the extinction of dinosaurs a cosmic accident or was their extinction planned? Why would their extinction be planned? Was the Earth being prepared for humans? On the other hand, is it possible that God placed fossils in the Earth to make the Earth appear old? Can you explain why God would create fossils?
3. Geologic Time: Suppose the lifetime of a human being is one hundred years and the Earth was formed 5 billion years ago. Can you use the lifetime of a human being to put into context the geologic time scale? Hint: the human being has 100 birthdays and lives approximately 36,500 days. A day in the life of a human being is like 137,000 years in the life of the Earth, and each year in the life of a human being is like 50 million years in the life of the Earth.
4. Science and Religion: How is it possible to believe in a supernatural being and in evolution?

Chapter 9. Mythological Creation Stories

1. Permanence of Current Beliefs: Many beliefs in the past are now considered myths, such as the gods and goddesses of ancient Egypt, Greece, and Rome. What beliefs today may be considered myths by future generations? Consider this question from the perspective of a believer, and from the perspective of a non-believer.
2. Science and Mythology: Over the centuries, scientists have offered various explanations of the origin of the universe and the origin of life. Would you characterize scientific explanations as myths? Explain your answer.
3. Do you believe in anything that someone else might consider a myth? Give an example.
4. Genesis and Mythology: Do you think the creation story in Genesis is a myth? Explain your answer.

Chapter 10. Origin of Religion

1. Religion and Society: What role does religion play in society? What would society be like without religion?
2. Religion and Society: Religion is subject to special rules in many societies. For example, some forms of religion may be outlawed or they may be accorded special tax treatment. Should atheism be considered a religion? Should it be treated as a religion by society?
3. Hell: Many people today do not believe in the existence of Satan/the devil/the Evil One or Hell. Do you agree? What would be the consequences if you were wrong?
4. Logic: A logic system such as mathematics is a system which obeys a set of rules. Do you believe everything in nature is subject to description by a logic system? Can you think of anything that might be illogical? For example, is behavior rational, where behavior includes emotional responses, such as love, hate, passion, lust, envy, faith, etc.?

Chapter 11. Eastern Theology

1. Internet Search: You can learn more about each of these religions on the internet. You can also find information about the number of adherents

to each religion and their geographic distribution. Why do you think people believe in these religions?
2. Which of these religions appeals to you most and why?
3. Mysticism: Have you had a mystical experience? How would you know?
4. Science and Religion: Many scientists believe that a theory of nature cannot be considered scientific unless it can be falsified, that is, shown to be incorrect. Can you find something in the teachings of any of these religions that would allow you to test the validity of the religion?

Chapter 12. Western Theology

1. Internet Search: You can learn more about each of these religions on the internet. You can also find information about the number of adherents to each religion and their geographic distribution. Why do you think people believe in these religions?
2. Which of these religions appeals to you most and why?
3. Love: Author C.S. Lewis defined four categories of love in his 1960 book *The Four Loves*: affection, friendship, romantic love (eros), and charity (also known as selfless love, Divine love, or agape). List some examples of love. How would you categorize your examples?
4. Science and Religion: Many scientists believe that a theory of nature cannot be considered scientific unless it can be falsified, that is, shown to be incorrect. Can you find something in the teachings of any of these religions that would allow you to test the validity of the religion?

Chapter 13. Beyond the Natural

1. Metaphysics and Science: The literal translation of metaphysics is "beyond physics" or "beyond the natural." Do you think anything is beyond physics or science in general? Explain your answer.
2. Reality: Is matter real? What do you mean by real?
3. Existence: Do you exist? What does existence mean to you?
4. Teleology: What is the meaning or purpose of the universe, if any?
5. Teleology: We can use science to study how the universe behaves. Is it appropriate to ask why the universe behaves as it does?

Chapter 14. Knowledge and Science

1. Reality of an Event: If an event occurs only once, how can it be tested and falsified? Examples of such events include Pascal's vision, the resurrection of Jesus, Mohammed speaking with an angel, and Moses' encounter with a burning bush. If the event cannot be tested and falsified, does that mean the event did not occur?
2. Reality of an Event: How many people have to witness an event before it is considered real? If only one person witnesses an event, does that person's knowledge of the event have any value?
3. Spiritualism: Have you had a spiritual experience? How would you know?
4. Science versus Religion: Some people believe that the conflict between science and religion is actually a conflict between the metaphysics of realism and the metaphysics of theism. Are the teachings of science and religion complementary or are they in conflict?

Chapter 15. Competing Worldviews?

1. Time and Genesis: According to the book of Genesis, God made the universe and everything in it in 6 days. How long is a day in Genesis? Is the word day symbolic of time duration, or is it a specific length of time? Relativity tells us that time duration depends on the observer. Who was the observer during the 6 days of creation?
2. Genesis 1 Interpretation: Does the creation account in Genesis 1 refer to the creation of matter or the creation of function? Does it matter which is correct?
3. Science and Religion: The relationship between science and religion has varied over the centuries and from one culture to another. What do you think the relationship is today? What do you think it should be?
4. Plato's Allegory of the Cave: Is the scientific worldview of a single universe a shadow of the reality of the Multi-Worlds Interpretation or multiverse? Can you relate Plato's allegory of the cave to our understanding of the multiverse?
5. Science and the Supernatural: Some scientists believe that there are many universes in a multiverse. Is it possible that there is a relationship

between the supernatural and a multiverse? Is scientific speculation about the multiverse more valid than belief in a supernatural?

Chapter 16. Theism or Atheism?

1. Risk: Do you like to gamble, or are you risk averse? What are some of the risks associated with Pascal's Wager?
2. Gambling and Risk: What kind of gambles have you made in your life? What risks have you taken? Did you feel compelled to make the gamble or take the risk? What were the consequences?
3. Nonbeliever's Wager: Suppose your answer to the question "does god exist?" is no. What happens when you die if you are right? What happens if you are wrong?
4. Believer's Wager: Suppose your answer to the question "does god exist?" is yes. What happens when you die if you are wrong? What happens if you are right?
5. Atheism and a Personal God: Suppose God exists, yet you choose to believe that god does not exist. Do you think God would forgive you?

Chapter 17. What Is Your Wager?

1. Free Will: What are some pivotal events in your life? How much control did you have over the event? Explain whether the event was inevitable, or if you had a choice?
2. Sin and Free Will: Schroeder [1997, p. 167] said the term "to sin" can be translated as "to err" or "to miss the mark." If we have free will, we are free to make choices. Sometimes our choices can cause harm even if we did not intend them to. Accidents can often be attributed to poor choices. Describe a choice you made that caused difficulties for someone else. Did you intend to cause harm?
3. Determinism and Free Will: Would it be possible for free will to exist if the universe was deterministic, that is, if everything that would ever happen in the universe was already known with certainty? Is our will free if the freedom was granted to us by an authority that had the power to repeal the freedom?

4. Good and Evil: Imagine there is no heaven, no hell, and no religion too (Yes, this is a parody of musician John Lennon's song). What should you use as a moral and ethical system of behavior? Should everyone use the same system? Why? Suppose someone chooses to abandon morality altogether. Do we have the right to judge that person, or should we tolerate behavior of all kinds? Do we have any authority to impose a system of morality on anyone else other than the authority of political power? Does might make right?
5. Spiritual Marker: There are several questions you can ask that can help you determine if an event could have been spiritual. Identify an event in your life that affected your lifestyle, such as a vision, graduation, choice of job, marriage, birth of a child, or death of a loved one. Answer the following questions: Did the event or decision impact your life or the lives of others? Did the event or decision improve your life or another life? How did you react to the event? Were you able to see all of the consequences of the event immediately, after much time had elapsed, or not at all? Do you believe that these events or decisions were inevitable or coincidental? Did they appear to establish a pattern that gave meaning to your life or helped guide your journey in life? Could the event have been a spiritual event? Hint: It may be easier to recognize spiritual markers in another person's life. Feel free to answer the above questions for someone you know.

Endnotes

Chapter 1

1. Genesis 1:1 [NIV *Bible*, 2002].
2. Exodus 20:2–17 and Deuteronomy 5:6–21 [NIV *Bible*, 2002].
3. Shahadah, or profession of faith, of Moslems; one of the Five Pillars of Islam [Smith, 1991].
4. Marx, K. [1843].
5. Nietzsche, F. [1882 and 1887].
6. Pais, A. [1982, especially p. 467] described Einstein's familiarity with Spinoza's work.
7. IARD Mission Statement, http://www.iard-relativity.org/, accessed November 15, 2020.
8. NOMA was introduced by Gould [1997] and developed further in Gould [2002].

Chapter 2

1. For more information about the cosmic ladder, see Silk [2001], Maoz [2007], Aughton [2008], Weinberg [2008], Bennett *et al.* [2018], and Lang [2018].
2. For more information about the history of the Big Bang, see Peebles [1993], Silk [2001], Singh [2004], Aughton [2008], Weinberg [2008], and Bennett *et al.* [2018].
3. Measurements of the cosmic microwave background (CMB) are discussed by Smoot and Scott [2000], Silk [2006], Bennett *et al.* [2018], and Fraknoi *et al.* [2018].
4. The Wilkinson Microwave Anisotropy Probe (WMAP) is discussed by Schwarzschild [2003; 2006], Silk [2006], Bennett *et al.* [2018], and Fraknoi *et al.* [2018].

Chapter 3

1. Sources on classical physics include D'Abro [1951], Bernal [1972], Jackson [1999], Morin [2007], and Leinaas [2019].
2. The literature on relativity is extensive. See, for example, Einstein [1905a; 1905b; 1916], Pauli [1958], Brouwer [1980], Ludvigsen [1999], Bernstein *et al.* [2000], Taylor *et al.* [2004], Smolin [2013], and Muller [2016]. See Jackson [1987; 1999] for a discussion of the impact of special relativity on theoretical physics. See Bernstein *et al.* [2000] for a discussion of the observational tests of special and general relativity.
3. American physicist John Archibald Wheeler [2010, p. 235] succinctly summarized the basic principle of general relativity: spacetime tells matter how to move; matter tells spacetime how to curve.
4. Some sources on cosmology include Einstein [1917], Weinberg [1993], Hogan [1998], Liddle [1999], Bernstein *et al.* [2000], Goldsmith [2000], Singh [2004], Tyson and Goldsmith [2004], Padmanabhan [2006], Silk [2006], Maoz [2007], Conselice [2007], Smolin [2013], Perlov and Vilenkin [2017], Bennett *et al.* [2018], Fraknoi *et al.* [2018], and Odenwald [2019].
5. For discussions of dark matter and dark energy, see Srednicki [2000], Bernstein *et al.* [2000], Perlmutter [2003], Tyson and Goldsmith [2004], Silk [2006], Gribbin [2006], Freeman and McNamara [2006], Schwarzschild [2007a, 2007b], Conselice [2007], Bennett *et al.* [2018], Fraknoi *et al.* [2018], and Odenwald [2019]. Rubin [2006] gives a personal account of one of the earliest observations that provided evidence for dark matter.
6. The scientific age of the earth was not determined without dispute. The development of a geological theory by Charles Lyell is documented by Gjertsen [1984], and the debate between physicist Lord Kelvin and members of the geology and biology communities is described by Hellman [1998]. Skinner [1986] reviewed evidence for the age of the earth and plate tectonics. Charles Darwin's role as a geologist is discussed by Herbert [1986].
7. See Pais [1982, Chapter 15e], Goldsmith [1995, p. 7], and Goldsmith [2000, Chapter 2].
8. Sources that discuss cosmological parameters include Misner *et al.* [1973], Peebles [1993], Weinberg [1993], Hogan [1998], Bergstrom and Goobar [1999], Liddle [1999], Ludvigsen [1999], Bernstein

et al. [2000], Fukugita and Hogan [2000], Goldsmith [2000], Kolb and Turner [2000], Singh [2004], Tyson and Goldsmith [2004], Padmanabhan [2006], Bennett *et al.* [2018], and Odenwald [2019]. A different approach to cosmology that does not rely on a Big Bang was advocated by Burbridge *et al.* [1999].

9. For discussions of inflation, see Linde [1987], Weinberg [1992], Peebles [1993], Krauss [1999], Bergstrom and Goobar [1999], Liddle [1999], Bucher and Spergel [1999], Kolb and Turner [2000], Singh [2004], Tyson and Goldsmith [2004], Silk [2006], and Padmanabhan [2006], Conselice [2007], Bennett *et al.* [2018], Fraknoi *et al.* [2018], and Odenwald [2019].
10. Stars and black holes are discussed in many sources, such as Misner *et al.* [1973], Peebles [1993], Blandford and Gehrels [1999], Padmanabhan [2006], Bennett *et al.* [2018], and Fraknoi *et al.* [2018].

Chapter 4

1. Sources on particle physics that include discussions of scattering, interactions, and fundamental forces include Close [1986], Perkins [1987], Cooper and West [1988], Bergstrom and Goobar [1999], Martin [2006], Huang [2007], Henley and Garcia [2007], and Odenwald [2019] The most recent summary of particle physics is published by the Particle Data Group [Zyla *et al.*, 2020] under the auspices of the United States Department of Energy.
2. The uncertainty principle is discussed by sources on quantum theory (see Endnote 3). Heisenberg's life has been chronicled in several sources, such as d'Abro [1951], Cassidy [1992], and Brennan [1997]. Heisenberg's relationship with Albert Einstein is discussed by Pais [1982] and Holton [2000].
3. Sources on quantum theory include Peebles [1992], Weinberg [1995], Bernstein *et al.* [2000], and Taylor *et al.* [2004].
4. For a review of Yukawa's work on meson theory, see Brown [1986].
5. Sources on relativistic quantum theory are numerous. A shortlist includes Bjorken and Drell [1964], Cooper and West [1988], Coughlin and Dodd [1991], Wu and Hwang [1991], Gross [1993], Greiner [1994], Weinberg [1995], Strange [1998], Bernstein *et al.* [2000], Zee [2003], and Huang [2007]. Hovis and Kragh [1993] present a short discussion of the life and views of P.A.M. Dirac, one of the founders

of relativistic quantum theory. For a critique of relativistic quantum mechanics, see Fanchi [1981; 1993]. A similar critique is presented in Weinberg [1995].

6. Several authors have speculated on the role of quantum fluctuations of the vacuum, including speculation on the creation of the cosmos from vacuum, quantum foams, and multiverses. For example, see Misner *et al.* [1973], Lindley [1987], Linde [1987], Liddle [1999], Greene [2000], Tyson and Goldsmith [2004], Silk [2006], Davies [2007], Hawking and Mlodinow [2010], Baggott [2013], Rovelli [2017], Bennett *et al.* [2018], Carroll [2019], and Odenwald [2019].
7. Discussions of theories beyond the Standard Model include Weinberg [1992], Hawking and Mlodinow [2010], Baggott [2013], Rovelli [2017], Carroll [2019], and Odenwald [2019].
8. String theory is discussed in several sources, such as Coughlan and Dodd [1991], Kaku [1993], Guidry [1999], Weinberg [1999], Greene [2000], Gates [2006], Carroll [2019], and Odenwald [2019]. Writers such as Woit [2006], Richter [2006], Smolin [2006], and Baggott [2013] questioned the scientific value of string theory. For example, Woit [2006] observed that part of the physics community studying superstring theory was beginning to minimize the importance of testable hypotheses. That same year, Smolin expressed concern about the level of funding going into superstring theory and how it was influencing career choices for young theorists [Smolin, 2006, Chapter 19].

Chapter 5

1. The origin and evolution of the universe and life are discussed by Kisslinger [2017], Teerikorpi *et al.* [2019], and Malkan and Zuckerman [2020].
2. Grand unified theories are discussed in such sources as Davies [1984], Zee [2003], Hawking and Mlodinow [2010], Baggott [2013], Rovelli [2017], Carroll [2019], and Odenwald [2019].
3. Theories beyond the Standard Model are discussed by Weinberg [1992], Hawking and Mlodinow [2010], Baggott [2013], Rovelli [2017], Carroll [2019], and Odenwald [2019].
4. Theories of nebular and galaxy formation are discussed by Hoyle [1963], Abel *et al.* [1991], Börner [1993], van den Bergh and Hesser [1993], Peebles [1993], Mather and Boslough [1996], Hester *et al.*

[2002], Silk [2006], Padmanabhan [2006], and Bennett *et al.* [2018]. The interstellar medium is discussed by Hester *et al.* [2002, Section 14.2], Reynolds [2002], Bennett *et al.* [2018], and Fraknoi *et al.* [2018]. The birth of stars is discussed by Hoyle [1963], Stahler [1991], Abel *et al.* [1991], Larson and Bromm [2001], Hester *et al.* [2002], Padmanabhan [2006], Bennett *et al.* [2018], and Fraknoi *et al.* [2018].

5. Binzel *et al.* [1991] discuss the origin of asteroids in the asteroid belt between Mars and Jupiter. Brown [2004], Bennett *et al.* [2018], and Fraknoi *et al.* [2018] discuss the Kuiper belt of objects beyond the orbit of Neptune. Some of the objects in the Kuiper belt are comparable in size to Pluto, which helped motivate members of the IAU to reconsider the meaning of the term planet and to reclassify Pluto as a dwarf planet in 2006.
6. Cosmic timelines have been presented by Weinberg [1977; 1993], Hogan [1998], Liddle [1999], Silk [2006], Bennett *et al.* [2018], Fraknoi *et al.* [2018], and OSUP [2020].

Chapter 6

1. A collection of background readings on the origin and evolution of life are introduced by Folsome [1979]. Early work on the origin of life issue is discussed by Oparin [1953], and Orgel [1973]. More recent discussions are presented by many authors, including Horowitz [1986], Shapiro [1987], Lahav [1999], Powell [2001], Tudge [2000b], Knoll [2003], Hazen [2005], Gribbin [2006], Rauchfuss [2008], Kisslinger [2017], Teerikorpi *et al.* [2019], and Malkan and Zuckerman [2020].
2. Sources on chemistry, biochemistry, and molecular biology include Orgel [1973], Shapiro [1987], Lahav [1999], Campbell *et al.* [1999], Purves *et al.* [2001], Tudge [2000b], Brown *et al.* [2003], Knoll [2003], Hazen [2005], Rauchfuss [2008], OSBio [2018], and OSChem [2019].
3. Advances in instrumentation and the ability to observe structures on the atomic level have facilitated the study of genes. See, for example, Keller [2000], Tudge [2000a], and Phillips and Quake [2006]. Bishop and Waldholtz [1999] describe the Human Genome Project which undertook the task of mapping all human genes. Collins [2006], who served as head of the Human Genome Project, makes a case that science and religion are complementary.

4. Many authors have proposed a definition for life. We consider the definitions discussed by Orgel [1973, p. 193], Horowitz [1986, p. 13], Lahav [1999, p. 113], Cohen [2000, p. 7], Hazen [2005, pp. 252–253], and Rauchfuss [2008, Section 1.4] as representative examples. Lahav [1999] presents several other definitions of life in his Appendix A (pp. 117–121).

Chapter 7

1. A collection of background readings on the origin and evolution of life are introduced by Folsome [1979]. Early work on the origin of life issue is discussed by Oparin [1953], and Orgel [1973]. More recent discussions are presented by many authors, including Horowitz [1986], Shapiro [1987], Lahav [1999], Powell [2001], Tudge [2000b], Knoll [2003], Hazen [2005], Gribbin [2006], Rauchfuss [2008], Kisslinger [2017], Schulze-Makuch and Irwin [2018], Teerikorpi *et al.* [2019], and Malkan and Zuckerman [2020].
2. Geological and geophysical sources include Siever [1983], Maxwell [1985], Skinner [1986], Montgomery [1990], Levin [1991], Ahrens [1994], Ridley [1996], Jeanloz and Romanowicz [1997], Fowler [2005], Reynolds *et al.* [2019], Grotzinger and Jordan [2020], and Lowrie [2020].
3. Discussions of radiometric dating (or isotopic dating) are provided in a variety of sources, including Montgomery [1990], Levin [1991], York [1993], Reynolds *et al.* [2019], Grotzinger and Jordan [2020], and Lowrie [2020] in geology and geophysics; Gould [1993] and Ridley [1996] in biology; and Williams [1991], Bernstein *et al.* [2000], and Lilley [2001] in physics.
4. Wegener's ideas were not readily accepted by proponents of the normal science of his time. For a discussion of the hurdles Wegener encountered, see Hellman [1998].
5. Sources on chemistry, biochemistry, and molecular biology include Orgel [1973], Shapiro [1987], Lahav [1999], Campbell *et al.* [1999], Purves *et al.* [2001], Tudge [2000b], Brown *et al.* [2003], Knoll [2003], Hazen [2005], Rauchfuss [2008], OSBio [2018], and OSChem [2019].
6. The role of climate in the origin of life is discussed by several authors, such as Cloud [1983], Horowitz [1986], Kasting *et al.* [1988], Gould [1993], Lahav [1999], Knoll [2003], Hazen [2005], Schulze-Makuch and Irwin [2018], and Teerikorpi *et al.* [2019].

7. For discussions of Panspermia and the Murchison meteorite, see Horowitz [1986], Lahav [1999], Hazen [2005], Schulze-Makuch and Irwin [2018], Fraknoi *et al.* [2018], and Teerikorpi *et al.* [2019]. For discussions of the Martian meteorite, see Ladbury [1996], Lahav [1999], and Knoll [2003]. The presence of water on Mars is discussed by Jakosky and Mellon [2004].
8. The Drake equation and Fermi Paradox are discussed by Schulze-Makuch and Irwin [2018, pp. 272–273], and Teerikorpi *et al.* [2019, pp. 544–545].
9. Sagan [1980] was a leading advocate of the belief that extraterrestrial life exists, and used Drake's equation to justify his position to argue that extraterrestrial life was highly likely. Other authors have argued that extraterrestrial life is improbable because it is difficult to find a location where the set of conditions that is needed to make life possible exists. These authors include Barrow and Tipler [1986], Ward and Brownlee [2000], Ross [2001, 2016], Knoll [2003], Schulze-Makuch and Irwin [2018], and Teerikorpi *et al.* [2019].
10. Cairns-Smith [1985a, 1985b], Gold [1999], Hazen [2001; 2005], Wächtershäuser [2006], and Schulze-Makuch and Irwin [2018] discuss theories that use minerals or extremophilic environments for the origins of life.

Chapter 8

1. A discussion of basic biology can be found in a variety of sources, such as Attenborough [1979], Wallace [1990], Campbell *et al.* [1999], Purves *et al.* [2001], and OSBio [2018].
2. References on evolution include Folsom [1979], Lewin [1982], Godfrey [1985], Gould [1993], Ridley [1996], Palmer [2009], and OSBio [2018]. Folsome, Godfrey, and Gould are compilations of essays on evolution. Most of these essays address modern variations of Darwinian evolution. Darwin's views can be obtained directly from his own writings [for reprints, see Darwin, 1952, 1959]. A case study of the evolution of an organism is given by Wong's [2002] description of the evolution of whales. Discussions of the controversial nature of Darwin's views can be found in many sources. The references above generally support Darwinian evolution. Johnson [1991] is an example of a book that challenges the validity of Darwinian evolution.

3. For a discussion of the relationship between progress and evolution, see Gould [1993, 1996]. The history and significance of the Burgess Shale are discussed by Gould [1989] and Knoll [2003].
4. An example of catastrophism is the catastrophist view of dinosaur extinction given by Russell [1982] and Alvarez [1987]. According to this view, a celestial body struck the earth and changed the climate enough to cause the extinction of many species of plants and animals. This view is supported by the observation of an impact crater of a comet or asteroid on the northern coast of the Yucatan peninsula. Gould [1993], Ridley [1996], and Gibbs [2001] review alternative theories for explaining mass extinctions, while Becker [2002] reviews several possible extinctions that may be attributable to collisions between the Earth and extraterrestrial objects. Mass extinction events are discussed by Benton [2003], Erwin [2006], OSBio [2018], Bennett *et al.* [2018], and Fraknoi *et al.* [2018].
5. Linnaeus' work was discussed by Gjertsen [1984]. Classification schemes that have replaced the two kingdoms approach were discussed by Whittaker [1969], Tudge [2000b], Palmer [2009], and OSBio [2018].
6. The taxonomy of Homo sapiens is still evolving. It is discussed by Gould [1993], Leakey [1994], Tudge [2000b], Palmer [2009], and OSBio [2018]. The geographic distribution of humans is discussed in several sources, including Tudge [2000b], Knoll [2003], Palmer [2009], and OSBio [2018].
7. There is a great deal of information about creationism and evolution in the literature and on the Internet. It is unwise to rely too heavily on characterizations of one group's position by another group. Given this disclaimer, a discussion of creationism by mainstream science proponents is given by Numbers [1982, 1992], Scott [1997], Pennock [1999], and Rennie [2002]. Internet sites for creationist organizations can be found with relative ease, and the organizations often have a link to a description of their history and beliefs.

Chapter 9

1. Sources that discuss Genesis include Van Over [1980], Armstrong [1993], Leeming and Leeming [1994], Bowker [2002], Walton [2009], and Harris and Platzner [2011].

2. Sources that discuss Egyptian mythology include Sproul [1979], Van Over [1980], Leeming and Leeming [1994], and Bowker [2002].
3. Sources that discuss the Enuma Elish include Van Over [1980], Armstrong [1993], and Bierlein [1994], and Bowker [2002].
4. Sources that discuss Greek mythology include Bierlein [1994], Souli [1995], Bowker [2002], and Harris and Platzner [2011].

Chapter 10

1. There is an extensive literature that discusses the origin and character of religion. See, for example, Berry [1956], Parrinder [1983], Smith [1965, 1991], Armstrong [1993], Bowker [2002], Barr [2003], Flew [2007], and Walton [2009].
2. Sources that discuss primitive (or ancient) religions include Berry [1956], Parrinder [1983], Armstrong [1993], and Walton [2009].
3. Suppes [1957], Carnap [1970], and Whorf [1987] were among the first researchers to note the dependence of higher-order thinking on language. Kuhn [1970], a philosopher of science, discussed the breakdown in communication between scientists with different perspectives when paradigms change.
4. Several authors have discussed the attitude of late nineteenth century physicists. For example, Weinberg [1992, pp. 13–15] discussed the feeling by some physicists that the major principles of physics were known by the end of the nineteenth century, but Horgan [1996, pp. 18–20] argued that this attitude may not have been as widely shared as the folklore of physics suggests.

Chapter 11

1. Capra [1975, 2010], Zukav [1979], and Talbot [1981] renewed interest in possible relationships between Eastern religions and concepts of modern physics and cosmology.
2. Interpretive references on religion were used in addition to original sources such as *Tao Te Ching* [Lao Tzu, 1955] and *The Bhagavad-Gita*. Sources on world religions include Berry [1956], Parrinder [1983], Smith [1965, 1991], Armstrong [1993], Reese [1999], Peterson *et al.*

[2001], Bowker [2002], and Barr [2003]. The lives of the founders of many of the world's major religions were dramatized by Vidal [1981].

Chapter 12

1. Interpretive references on religion were used in addition to original sources such as the *Bible* and the *Koran*. Sources on world religions include Berry [1956], Parrinder [1983], Smith [1965, 1991], Armstrong [1993], Reese [1999], Peterson *et al.* [2001], Bowker [2002], and Barr [2003]. The lives of the founders of many of the world's major religions were dramatized by Vidal [1981].
2. The word "man" is used in the historical biblical sense as a representation for humanity or human beings. Context indicates when "man" refers to a specific gender.
3. Reviews of ancient evidence and arguments for and against the existence of Jesus are discussed by such authors as Van Voorst [2000] Lightfoot [2003], Jones [2015], and Kennedy [2020].
4. *Koran* is the Romanized spelling of Quran.

Chapter 13

1. Much of the discussion of philosophical concepts is based on translations of the original papers, many of which are collected in Stroll and Popkins [1972]. Hick [1970], Russell [1972], Reese [1980], Barrow and Tipler [1986], Cushing [1998], Peterson *et al.* [2001], and Cronk *et al.* [2004] served as supplementary references.
2. Pais [1982, especially p. 467] noted Einstein's familiarity with Spinoza's work, and Calder [1982, p. 138] repeated a telegram that Einstein sent to a Jewish newspaper in 1929 that said "I believe in Spinoza's God who reveals himself in the harmony of all that exists, but not in a God who concerns himself with the fate and action of men."

Chapter 14

1. Much of the discussion of philosophical concepts is based on translations of the original papers, many of which are collected in Stroll and Popkins [1972]. Hick [1970], Russell [1972], Reese [1980],

Barrow and Tipler [1986], Cushing [1998], Peterson *et al.* [2001], and Cronk *et al.* [2004] served as supplementary references.

Chapter 15

1. The conflict between science and religion may be viewed as a conflict between realism and theism [Barr, 2003]. The impact of science and religion on cosmology is discussed by Kragh [2004].
2. Several authors have discussed multiverses. For example, see Lindley [1987], Tyson and Goldsmith [2004], Silk [2006], Davies [2007], Hawking and Mlodinow [2010], Krauss [2012], Baggott [2013], Aczel [2015], Perlov and Vilenkin [2017], Fraknoi *et al.* [2018], Carroll [2019], and Odenwald [2019].
3. Some sources on astronomy include Hester *et al.* [20002], Aughton (2008), Lang [2018], Bennett *et al.* [2018], Fraknoi *et al.* [2018], and Odenwald [2019].
4. Sources on world religions include Parrinder [1983], Smith [1965, 1991], Berry [1956], Reese [1999], Peterson *et al.* [2001], and Bowker [2002].

Chapter 16

1. Hick [1970], Reese [1999], Peterson *et al.* [2001], and Flew [2007] present succinct discussions of the arguments for the existence of God. Hick and Peterson *et al.* provide a collection of related readings in the philosophy of religion. Dawkins [2006] and Krauss [2012] present perspectives on atheism.
2. Flew [2007] included a dialogue on Jesus with N.T. Wright in Appendix B to illustrate arguments that influenced Flew's conversion to theism.
3. The anthropic principle has been discussed by Barrow and Tipler [1986], Weinberg [1992], Peebles [1993], Hogan [1998], Ward and Brownlee [2000], Tyson and Goldsmith [2004], Silk [2006], Benedict [2013], Aczel [2015], Perlov and Vilenkin [2017], Hawking [2018], and Teerikorpi *et al.* [2019]. The anthropic principle is not universally accepted.
4. Feuerbach is quoted in Stroll and Popkins [1972]. Philosophical treatises on atheism, especially by Feuerbach and Kierkegaard, are

presented in Hick [1970]. Various aspects of atheism are discussed by Armstrong [1993], Dawkins [2006], and Hitchens [2008].

5. More information about humanism is available from the American Humanist Association at https://americanhumanist.org. Dewey's views are summarized in a number of sources, including Hick [1970], Reese [1980], Stallman [2003], and Woolfolk [2019].
6. Biographer James Connor [2006] presented a translation of the note Pascal wrote on 23 November 1654 describing the mystical experience.

Chapter 17

1. For example, see religion philosopher Paul Brockelman [1999, p. 177]; see also physicist-astronomer Chet Raymo [1998]. By contrast, scientists like Brian Greene [2020] argue that humans must provide our own purpose.

References

Abel, G.O., D. Morrison, and S.C. Wolff (1991). *Exploration of the Universe*, 6th Edition, Saunders, Philadelphia.

Aczel, A. (2014). *Why Science Does Not Disprove God*, William Morrow, HarperCollins, New York.

Ahrens, T.J. (1994). The Origin of the Earth, *Physics Today* (August), pp. 38–45.

Alvarez, L.W. (1987). Mass Extinctions Caused by Large Bolide Impacts, *Physics Today* (July), pp. 24–33.

Amlin, P. (1989). *Popol Vuh: The Creation Myth of the Maya*, English language version, National Endowment for the Humanities, Washington, D.C.

Anderson, P.W. (1972). More is Different, *Science*, Volume 147, pp. 393–396.

Armstrong, K. (1993). *A History of God*, Ballantine Books, Random House, New York.

ATNF (2021). The Death of Stars I: Solar-Mass Stars, Sun's Post-Main Sequence Evolutionary Track, Australia Telescope National Facility, accessed February 5, 2021 from https://www.atnf.csiro.au/outreach//education/senior/astrophysics/stellarevolution_deathlow.html.

Aughton, P. (2008). *The Story of Astronomy*, Quercus, London.

Augustine of Hippo (ca. 400 CE). *Confessions of Saint Augustine*, Franklin Library, Franklin Center, Pennsylvania; printed in 1976.

Bada, J.L. and A. Lazcano (2012). Stanley L. Miller — A Biographical Memoir, United States National Academy of Sciences, accessed December 30, 2020 from http://nasonline.org/publications/biographical-memoirs/memoir-pdfs/miller-stanley.pdf.

Baggott, J. (2013). *Farewell to Reality*, Pegasus Books, London.

Barr, S.M. (2003). *Modern Physics and Ancient Faith*, University of Notre Dame Press, Notre Dame, Indiana.

Barrow, J.D. and F.J. Tipler (1986). *The Anthropic Cosmological Principle*, Oxford University Press, New York.

Becker, L. (2002). Repeated Blows, *Scientific American* (March), pp. 77–83.

Benedict, G. (2013). *The God Debate*, Watkins Publishing, London.
Bennett, J., M. Donahue, N. Schneider, and M. Voit (2018). *The Essential Cosmic Perspective*, 8th Edition, Pearson, New York.
Benton, M.J. (2003). *When Life Nearly Died*, Thames & Hudson, London.
Bergstrom, L. and A. Goobar (1999). *Cosmology and Particle Astrophysics*, Wiley, New York.
Bernal, J.D. (1972). *The History of Classical Physics*, published in 1999 by Barnes & Noble, New York.
Bernstein, A. (1986). Thinking About Hell, *The Wilson Quarterly*, Summer issue, The Wilson Center, Washington, D.C.
Bernstein, J., P.M. Fishbane, and S. Gasiorowicz (2000). *Modern Physics*, Prentice Hall, Upper Saddle River, New Jersey.
Berry, G.L. (1956). *Religions of the World*, Barnes and Noble, New York.
Bhagavad-Gita (1972). Translated by Swami Prabhavananda and C. Isherwood. New American Library, New York.
Bible, NIV (2002). *New International Version Bible*, Zondervan, Grand Rapids, Michigan.
Bible, RSV (1971). *Revised Standard Version Bible, American Bible Society*, New York.
Binzel, R.P., M.A. Barucci, and M. Fulchignoni (1991). The Origins of the Asteroids, *Scientific American*, October, pp. 88–94.
Bishop, J.E. and M. Waldholtz (1999). Genome, iUniverse.com, Lincoln, Nebraska.
Bierlein, J.F. (1994). *Parallel Myths*, Ballantine, New York.
Bjerknes, C.J. (2003). *Anticipations of Einstein in the General Theory of Relativity*, XTX, Downers Grove, Illinois.
Bjorken, J.D. and S.D. Drell (1964). *Relativistic Quantum Mechanics*, McGraw-Hill, New York.
Black Smoker (2021). Black Smoker image accessed January 2, 2021 from http://earthsci.org/mineral/energy/geomin/geomin.html.
Blandford, R. and N. Gehrels (1999). Revisiting the Black Hole, *Physics Today* (June), pp. 40–46.
Bondi, H. and T. Gold (1948). The Steady-State Theory of the Expanding Universe, *MNRAS*, Volume 108, pp. 252–270.
Börner, G. (1993). *The Early Universe*, Springer-Verlag, Berlin.

Bowker, J. (2002). *God — A Brief History*, DK Publishing, London.
Brennan, R.P. (1997). *Heisenberg Probably Slept Here*, Wiley, New York.
Brockelman, P. (1999). *Cosmology and Creation*, Oxford University Press, New York.
de Broglie, L. (1966). *Physics and Microphysics*, Grosset's Universal Library, Grosset and Dunlap, New York.
Brouwer, W. (1980). Einstein and Lorentz: The Structure of a Scientific Revolution, *American Journal of Physics*, Volume 48, pp. 425–431.
Brown, D. (2000). *Angels & Demons*, Simon & Schuster, New York.
Brown, L.M. (1986). Hideki Yukawa and the Meson Theory, *Physics Today* (December), pp. 55–62.
Brown, M.E. (2004). The Kuiper Belt, *Physics Today* (April), pp. 49–54.
Brown, T.L., H.E. LeMay, Jr., B.E. Bursten, J.R. Burdge (2003). *Chemistry — The Central Science*, 9th Edition, Prentice Hall, Upper Saddle River, New Jersey.
Bucher, M.A. and D.N. Spergel (1999). Inflation in a Low-Density Universe, *Scientific American* (January), pp. 63–69.
Burbridge, G., F. Hoyle, and J.V. Narlikar (1999). A Different Approach to Cosmology, *Physics Today*, Volume 52, pp. 38–44. A reply by A. Albrecht representing normal science is appended to the article, *ibid.*, pp. 44–46.
Byrne, P. (2007). The Many Worlds of Hugh Everett, *Scientific American* (December), Volume 297, Issue 6, pp. 98–105.
Cairns-Smith, A.G. (1985a). *Seven Clues to the Origin of Life*, Cambridge University Press, Cambridge, United Kingdom.
Cairns-Smith, A.G. (1985b). The First Organisms, *Scientific American* (June), pp. 90–100.
Calder, N. (1982). *Einstein's Universe*, Greenwich House, New York.
Campbell, N.A., J.B. Reece, and L.G. Mitchell (1999). *Biology*, 5th Edition, Benjamin-Cummins, Menlo Park, California.
Capra, F. (1975). *The Tao of Physics*, Bantam, New York.
Capra, F. (2010). *The Tao of Physics*, 5th Edition. Shambhala Publications, Boston.
Carnap, R. (1970). *Foundations of Logic and Mathematics*, University of Chicago Press, Chicago.
Carroll, S. (2012). *The Particle at the End of the Universe*, Penguin Group, New York.

Carroll, S. (2019). *Something Deeply Hidden*, Dutton-Penguin Random House, New York.

Cassidy, D.C. (1992). Heisenberg, Uncertainty and the Quantum Revolution, *Scientific American* (May), pp. 106–112.

de Chardin, Teilhard (1975): *The Phenomenon of Man*, first published in French in 1955, translated by B. Wall, Harper & Row, New York.

Close, F. (1986). *The Cosmic Onion*, American Institute of Physics, New York.

Cloud, P. (1983). The Biosphere, *Scientific American* (September), pp. 176–189.

CMS Collaboration (2012). Observation of a New Boson at a Mass of 125 GeV with the CMS Experiment at the LHC, *Physics Letters B*, Volume 716, pp. 30–61.

Cohen, R. (2000). *History of Life*, 3rd Edition, Blackwell Science, Malden, Massachusetts.

Collins, F.S. (2006). *The Language of God*, Free Press, Simon & Schuster, New York.

Connor, J.A. (2006). *Pascal's Wager*, Harper Collins, San Francisco.

Conselice, C.J. (2007). The Universe's Invisible Hand, *Scientific American* (February), pp. 34–41.

Cooper, N.G. and G.B. West (1988). *Particle Physics: A Los Alamos Primer*, Cambridge University Press, Cambridge, United Kingdom.

Coughlan, G.D. and J.E. Dodd (1991). *The Ideas of Particle Physics*, 2nd Edition, Cambridge University Press, Cambridge, United Kingdom.

Cronk, G., T. de Marco, P. Dlugos and P. Eckstein (2004). *Readings in Philosophy*, 2nd Edition, Hayden McNeil Publishing, Plymouth, Michigan.

Cushing, J.T. (1998). *Philosophical Concepts in Physics*, Cambridge University Press, Cambridge, United Kingdom.

D'Abro, A. (1951). *The Rise of the New Physics*, Dover, New York, 2 volumes.

Darwin, C. (1952). *The Origin of Species and The Descent of Man*, Great Books Volume 49, Encyclopedia Britannica, Chicago.

Darwin, C. (1959). *The Voyage of the Beagle*, Harper and Brothers, New York.

Darwin, E. (1794). Zoonomia, Byrne and Jones, Dublin; accessed January 4, 2021 from https://early-evolution.oeb.harvard.edu/files/early-evolution/files/zoonomia.pdf?m=1432129930.

Davies, P.C.W. (1984). *Superforce*, Simon and Schuster, New York.

Davies, P. (2007). Universes Galore: Where Will it all End? In B. Carr (Ed.), *Universe or Multiverse?* pp. 487–506, Cambridge University Press, Cambridge, United Kingdom, doi:10.1017/CBO9781107050990.030.

Dawkins, R. (2006). *The God Delusion*, Houghton Mifflin, Boston.

Durkheim, É. (2001). *The Elementary Forms of Religious Life*, first published in 1912, translated from the 1991 Livre de Poche French edition, Oxford University Press, Oxford, United Kingdom.

Ecklund, E.H., D.R. Johnson, C.P. Scheitle, K.R.W. Matthews, and S.W. Lewis (2016). Religion among Scientists in International Context: A New Study of Scientists in Eight Regions, *Socius: Sociological Research for a Dynamic World*, Volume 2, pp. 1–9.

Eddington, A.S. (1923). *The Mathematical Theory of Relativity*, 2nd Edition in 1924, reprinted in 1960 by Cambridge University Press, Cambridge, United Kingdom.

Einstein, A. (1905a). On the Electrodynamics of Moving Bodies, first published in *Annalen der Physik*, Volume 17; English translation first published in The Principle of Relativity, 1923, Dover, New York, pp. 35–66.

Einstein, A. (1905b). Does the Inertia of a Body Depend Upon its Energy-Content?, first published in *Annalen der Physik*, Volume 17; English translation first published in The Principle of Relativity, 1923, Dover, New York, pp. 67–71.

Einstein, A. (1916). The Foundation of the General Theory of Relativity, first published in *Annalen der Physik*, Volume 49, pp. 769–822; English translation published in The Principle of Relativity, Methuen, 1923, reprinted by Dover, New York, pp. 109–164.

Einstein, A. (1917). *Cosmological Considerations of the General Theory of Relativity*, first published in Sitzungsberichte der Preussischen Akad. D. Wissenschafte; English translation published in The Principle of Relativity, Methuen,1923, reprinted by Dover, New York, pp. 175–188.

Fanchi, J.R. (1977). Quantum Mechanics of Relativistic Spinless Particles, Ph.D. Dissertation, University of Houston.

Fanchi, J.R. (1981). A Critique of Conventional Relativistic Quantum Mechanics, *American Journal of Physics*, Volume 49, pp. 851–853.

Fanchi, J.R. (1988). Cosmological Implications of the Gibbs Ensemble in Parametrized Relativistic Classical Mechanics, *Physical Review A*, Volume 37, May, pp. 3956–3962.

Fanchi, J.R. (1993). *Parametrized Relativistic Quantum Theory*, Kluwer, Dordrecht, Netherlands.

Fanchi, J.R. (2002). *Shared Earth Modeling*, Butterworth-Heinemann, Boston, Massachusetts.

Fanchi, J.R. (2004). *Energy: Technology and Directions for the Future*, Elsevier Academic, Burlington, Massachusetts.

Fanchi, J.R. (2018). *Principles of Applied Reservoir Simulation*, 4th Edition, Gulf Professional Publishing, Elsevier, Cambridge, Massachusetts.

Flew, A. (2007). *There is a God*, with Preface and Appendix by Roy Abraham Varghese, HarperCollins Publishers, New York.

Folsome, C.E. (1979). *Life: Origin and Evolution, Readings from Scientific American*, W.H. Freeman, San Francisco.

Fowler, C.M.R. (2005). *The Solid Earth*, 2nd Edition, Cambridge University Press, Cambridge, United Kingdom.

Fox, S.W. (1997). My Scientific Discussions of Evolution for the Pope and His Scientists, Harbinger Symposium, Introduction by Sheldon Gottlieb, accessed December 30, 2020 from http://www.theharbinger.org/articles/rel_sci/fox.html.

Fraknoi, A., D. Morrison, and S.C. Wolff (2018). Astronomy OpenStax, Rice University, Houston, accessed December 23, 2020 from https://openstax.org/details/books/astronomy.

Frazier, J.G. (1890). *The Golden Bough: A Comparative Study of Religion*, published in 2 volumes in1894 by MacMillan, New York and London; accessed January 27, 2021 from the Internet Archive at https://archive.org.

Freeman, K. and G. McNamara (2006). *In Search of Dark Matter*, Springer-Praxis, Chichester, United Kingdom.

Friedlander, M.W. (1995). *At the Fringes of Science*, Westview Press, Boulder, Colorado.

Fukugita, M. and C.J. Hogan (2000). Global Cosmological Parameters, *The European Physical Journal C*, Volume 15, pp. 136–142.

Galilei, Galileo (1615). Letter to the Grand Duchess Christina of Tuscany (abridged), accessed February 5, 2021 from the Internet Modern History Sourcebook at https://web.stanford.edu/~jsabol/certainty/readings/Galileo-LetterDuchessChristina.pdf.

Gates, S.J. Jr. (2006). Is String Theory Phenomenologically Viable?, *Scientific American* (June), pp. 54–59.

Gibbs, W.W. (2001). On the Termination of Species, *Scientific American* (November), pp. 40–49.
Gjertsen, D. (1984). *The Classics of Science*, Lilian Barber Press, New York.
Glashow, S.L. (1980). Towards a Unified Theory: Threads in a Tapestry, *Reviews of Modern Physics*, Volume 52, pp. 539 ff.
Godfrey, L.R. (1985). *What Darwin Began*, Allyn and Bacon, Newton, Massachusetts.
Gold, T. (1999). *The Deep Hot Biosphere*, Springer-Verlag, New York.
Goldsmith, D. (1995). *Einstein's Greatest Blunder*, Harvard University Press, Cambridge, Massachusetts.
Goldsmith, D. (2000). *The Runaway Universe*, Perseus Books, Cambridge, Massachusetts.
Gould, S.J. (1989). *Wonderful Life*, Norton, New York.
Gould, S.J., editor (1993). *The Book of Life*, W.W. Norton, New York.
Gould, S.J. (1996). *Full House*, Harmony Books, New York.
Gould, S.J. (1997). Nonoverlapping Magisteria, *Natural History Magazine*, Volume 106 (March), pp. 16–22.
Gould, S.J. (2002). *Rocks of Ages: Science and Religion in the Fullness of Life*, Ballantine Books, New York.
Greene, B. (2000). *The Elegant Universe*, Vintage Books, New York.
Greene, B. (2020). *Until the End of Time*, Penguin Random House, New York.
Greiner, W. (1994). *Relativistic Quantum Mechanics*, Springer-Verlag, Berlin.
Gribbin, J. (2006). *The Origins of the Future*, Yale University Press, New Haven, Connecticut.
Gross, F. (1993). *Relativistic Quantum Mechanics and Field Theory*, Wiley, New York.
Grotzinger, J. and T.H. Jordan (2020). *Understanding Earth*, 8th Edition, W.H. Freeman and Company, New York.
Guidry, M. (1999). *Gauge Field Theories*, Wiley, New York.
Guth, A. (2002). Inflation and the New Era of High-Precision Cosmology, *MIT Physics Annual*, MIT, Cambridge, Massachusetts; accessed December 15, 2020 from http://web.mit.edu/physics/news/physicsatmit/physicsatmit_02_cosmology.pdf.
Hamilton, E. (1942). *Mythology*, Little, Brown and Company, Boston.
Harris, S. and G. Platzner (2011). *Classical Mythology*, 6th Edition, McGraw-Hill, Boston.
Hawking, S. (2001). *The Universe in a Nutshell*, Bantam Books, New York.

Hawking, S. (2018). *Brief Answers to the Big Questions*, Bantam Books, New York.

Hawking, S. and L. Mlodinow (2010). *The Grand Design*, Bantam Books, New York.

Hazen, R.M. (2005). *Genesis: The Scientific Quest for Life's Origins*, Joseph Henry Press, Washington, D.C.

Heisenberg, W. (1927). Über den anschaulichen Inhalt der quantentheoretischen Kinematik und Mechanik, *Zeitschrift für Physik*, Volume 43, pp. 172–198; English translation (The actual content of quantum theoretical kinematics and mechanics) by NASA in 1983, accessed December 3, 2020 from https://ia800500.us.archive.org/10/items/nasa_techdoc_19840008978/19840008978.pdf.

Hellman, H. (1998). *Great Feuds in Science*, Wiley, New York.

Henley, E.M., and A. Garcia (2007). *Subatomic Particles*, 3rd Edition, World Scientific, Singapore.

Herbert, S. (1986). Darwin as a Geologist, *Scientific American* (May), pp. 116–123.

Hester, J., D. Burstein, G. Blumenthal, R. Greeley, B. Smith, H. Voss, and G. Wegner (2002). *21st Century Astronomy*, W.W. Norton, New York.

Hick, J. (1970). *Philosophy of Religion*, 2nd Edition, Prentice-Hall, Englewood Cliffs, New Jersey.

Hitchens, C. (2008). *The Portable Atheist: Essential Readings for the Nonbeliever*, Perseus Publishing, New York.

Hodgkinson, B. (2006). *The Essence of Vedanta*, Chartwell Books, Edison, New Jersey.

Hogan, C.J. (1998). *The Little Book of the Big Bang*, Springer-Verlag, New York.

Holton, G. (2000). Werner Heisenberg and Albert Einstein, *Physics Today* (July), pp. 38–42.

Horgan, J. (1996). *The End of Science*, Addison-Wesley, Reading, Massachusetts.

Horowitz, N. (1986). *To Utopia and Back*, W.H. Freeman, New York.

Hovis, R.C. and H. Kragh (1993). P.A.M. Dirac and the Beauty of Physics, *Scientific American* (May), pp. 104–109.

Hoyle, F. (1948). A New Model for the Expanding Universe, *MNRAS*, Volume 108, pp. 372–382.

Hoyle, F. (1963). *Frontiers of Astronomy*, reprint of 1955 edition by Signet Science Library, Harper & Row, New York.

Huang, K. (2007). *Fundamental Forces of Nature*, World Scientific, Singapore.

Hubble, E. (1929). A Relation Between Distance and Radial Velocity Among Extra-Galactic Nebula, *Proceedings of the National Academy of Sciences*, Volume 15, National Academy of Sciences, Washington, D.C., pp. 168–173.

Hubble, E. and M.L. Humason (1931). The Velocity-Distance Relation among Extra-Galactic Nebulae, *Astrophysical Journal*, Volume 74, pp. 43–80.

Jackson, J.D. (1987). The Impact of Special Relativity on Theoretical Physics, *Physics Today* (May), pp. 34–42.

Jackson, J.D. (1999). *Classical Electrodynamics*, 3rd Edition, Wiley, New York.

Jakosky, B.M. and M.T. Mellon (2004). Water on Mars, *Physics Today* (April), pp. 71–76.

Jeanloz, R. and B. Romanowicz (1997). Geophysical Dynamics at the Center of the Earth, *Physics Today* (August), pp. 22–27.

Johnson, P.E. (1991). *Darwin on Trial*, Regnery Gateway, Washington, D.C.

Jones, T.P. (2015). *How We Got the Bible*, Rose Publishing, Torrance, California.

Josephus, Flavius (ca. 93 CE). *Book XVII, Antiquities of the Jews, The Life and Essential Works of Flavius Josephus*, translated in 1737 by W. Whiston, published in 2013 by Plain Sight Publishing, Springville, Utah.

Kaku, M. (1993). *Quantum Field Theory*, Oxford University Press, Oxford, United Kingdom.

Kasting, J.F., O.B. Toon, and J.B. Pollack (1988). How Climate Evolved on the Terrestrial Planets, *Scientific American* (February), pp. 90–97.

Keller, E.F. (2000). *The Century of the Gene*, Harvard University Press, Cambridge, Massachusetts.

Kennedy, T.M. (2020). *Unearthing the Bible*, Harvest House Published, Eugene, Oregon.

Kisslinger, L.S. (2017). *Astrophysics and the Evolution of the Universe*, 2nd Edition, World Scientific, Singapore.

Kline, A.S. (2000). *Dante — The Divine Comedy*, authored by Dante Alighieri between 1308 and his death in 1321, prose translation by A.S. Kline, illustrated by G. Dore, accessed February 7, 2021 from https://www.poetryintranslation.com/klineasdante.php.

Knoll, A.H. (2003). *Life on a Young Planet*, Princeton University Press, Princeton, New Jersey.

Kolb, E.W. and M.S. Turner (2000). The Pocket Cosmology, *The European Physical Journal C*, Volume 15, pp. 125–132.

Koran (1974). Translated by N.J. Dawood, Penguin Classics, New York.

Koran (1996). Translated by N.J. Dawood, Easton Press, Norwalk, Connecticut.

Kragh, H. (2004). *Matter and Spirit in the Universe — Scientific and Religious Preludes to Modern Cosmology*, Imperial College Press, London.

Kramer, F. (1970). *Rock Strata and the Bible Record*, edited by P.A. Zimmerman, Concordia Publishing House, St. Louis.

Krauss, L.M. (1999). Cosmological Antigravity, *Scientific American* (January), pp. 53–59.

Krauss, L. (2012). *A Universe from Nothing*, Simon & Schuster, New York.

Kuhn, T.S. (1970). *The Structure of Scientific Revolutions*, 2nd Edition, University of Chicago Press, Chicago.

Ladbury, R. (1996). Martians Invaded Earth 13,000 Years Ago — Maybe, *Physics Today* (September), p. 18.

Lahav, N. (1999). *Biogenesis*, Oxford University Press, Oxford, United Kingdom.

Lamont, C. (1990). *The Philosophy of Humanism*, Continuum, New York.

Lang, K.R. (2018). *A Brief History of Astronomy and Astrophysics*, World Scientific, Singapore.

Lao Tzu (1955). *The Way of Life (aka Tao Te Ching)*, translated by R.B. Blakney. New American Library, New York.

Larson, R.B. and V. Bromm (2001). The First Stars in the Universe, *Scientific American* (December), pp. 54–71.

Larson, E.J. and L. Witham (1998). Leading Scientists Still Reject God, *Nature Correspondence*, Volume 394, 23 July 1998, p. 313.

Leavitt, H.S. (1908). 1777 Variables in the Magellanic Clouds, *Annals of Harvard College Observatory*, Volume 60, Number 4, pp. 87–108.3.

Leavitt, H.S. and E.C. Pickering (1912). Periods of 25 Variable Stars in the Small Magellanic Cloud, *Harvard College Observatory Circular*, Volume 173, March 3, 1912, pp. 1–3; accessed December 24, 2020 from https://silentskyplay.tumblr.com/post/71219744368/henrietta-swan-leavitts-actual-1912-paper-wherein.

Leeming, D. and M. Leeming (1994). *A Dictionary of Creation Myths*, Oxford University Press, New York.

Leinaas, J.M. (2019). *Classical Mechanics and Electrodynamics*, World Scientific, Singapore.

Levin, H.L. (1991). *The Earth Through Time*, 4th Edition, Harcourt-Brace Jovanovich, New York.

Lewis, C.S. (1952). *Mere Christianity*, HarperCollins e-books.

Liddle, A. (1999). *An Introduction to Modern Cosmology*, Wiley, New York.

Lightfoot, N.R. (2003). *How We Got the Bible*, 3rd Edition Revised and Expanded, MJF Books, New York.

Lilley, J. (2001). *Nuclear Physics*, Wiley, New York.

Linde, A. (1987). Particle Physics and Inflationary Cosmology, *Physics Today* (September issue), pp. 61–68.

Lindley, D. (1987). Cosmology from Nothing, *Nature* (December), pp. 603–604.

Lowrie, W. (2020). *Fundamentals of Geophysics*, Cambridge University Press, 3rd Edition, Cambridge, United Kingdom.

Lucado, M. (2007). *3:16 The Numbers of Hope*, Easton Press, Norwalk, Connecticut.

Ludvigsen, M. (1999). *General Relativity*, Cambridge University Press, Cambridge, United Kingdom.

Lyell, C. (1830–1833). *Principles of Geology*, Volumes I through III, Gryphon Editions; New York; printed in 1997.

Malkan, M.A. and B. Zuckerman, editors (2020). *Origin and Evolution of the Universe*, 2nd Edition, World Scientific, Singapore.

Maoz, D. (2007). *Astrophysics in a Nutshell*, Princeton University Press, Princeton, New Jersey.

Marett, R.R. (1900). Pre-Animistic Religion, *Folklore*, Volume 11, Issue 2, pp. 162–184; first read as a talk in 1899; accessed January 27, 2021 from http://www.jstor.org/stable/1253250.

Marin, J.M. (2009). 'Mysticism' in Quantum Mechanics: The Forgotten Controversy, *European Journal of Physics*, Volume 30, pp. 807–822.

Martin, B.R. (2006). *Nuclear and Particle Physics*, John Wiley & Sons, New York.

Marx, K. (1843). *Critique of Hegel's Philosophy of Right*, reprinted in 1977 by Cambridge University Press, Cambridge, United Kingdom.

Mather, J.C. and J. Boslough (1996). *The Very First Light*, Basic Books, New York.

Maxwell, J.C. (1985). What is the lithosphere?, *Physics Today* (September). pp. 32–40.

Mayr, Ernst (1982). *The Growth of Biological Thought*, Cambridge, Massachusetts; Belknap Press of Harvard University Press.

McKay, C.P. (2020). The Origin and Evolution of Life in the Universe, Chapter 6 in *Origin and Evolution of the Universe*, 2nd Edition, Malkan, M.A. and B. Zuckerman, editors, World Scientific, Singapore.

Miller, S.L. (1953). A Production of Organic Compounds under Possible Primitive Earth Conditions, *Science*, Volume 117, pp. 528–529.

Milton, J. (1667). *Paradise Lost*, Easton Press, Norwalk, Connecticut; printed in 1976.

Misner, C.W., K.S. Thorne, and J.A. Wheeler (1973). *Gravitation*, W.H. Freeman, San Francisco.

Montgomery, C.W. (1990). *Physical Geology*, 2nd Edition, Wm. C. Brown, Dubuque, Iowa.

More, T. (1516). *Utopia*, first published in 1516, translated by Ralph Robynson (ca. 1551); printed in 1963 by Easton Press, Norwalk, Connecticut.

Morin, D. (2007). *Introduction to Classical Mechanics*, Cambridge University Press, Cambridge, United Kingdom.

Muller, R.A. (2016). *Now: The Physics of Time*, W.W. Norton and Co., New York.

NAS (1998). *Teaching About Evolution and the Nature of Science*, Members of the National Academy of Sciences, National Academy Press, Washington, D.C.

NASA-COBE (2009). *COBE Satellite marks 20th Anniversary*, published online November 17, 2009 by the National Aeronautics and Space Administration, accessed October 13, 2019 from https://www.nasa.gov/topics/universe/features/cobe_20th.html.

NASA Exobiology (2020). Exobiology, National Aeronautics and Space Administration, accessed December 28, 2020 from https://astrobiology.nasa.gov/research/astrobiology-at-nasa/exobiology/.

NASA Geometry (2020). Geometry of the Universe, National Aeronautics and Space Administration, accessed December 8, 2020 from https://map.gsfc.nasa.gov/media/990006/index.html.

NASA-WMAP (2010). *NASA's WMAP Project Completes Satellite Operations*, Published online October 6, 2010 by the National Aeronautics and Space Administration WMAP Science Team, accessed November

27, 2020 from https://www.nasa.gov/topics/universe/features/wmap-complete.html,seealsohttps://wmap.gsfc.nasa.gov/universe/bb_cosmo_fluct.html.
NASA-WMAP Science Team (2011). Cosmology: The Study of the Universe, National Aeronautics and Space Administration, Wilkinson Microwave Anisotropy Probe, last modified June 6, 2011, http://map.gsfc.nasa.gov/universe/WMAP_Universe.pdf or http://map.gsfc.nasa.gov/universe/.
Newton, I. (1687). *Philosophiae Naturalis Principia Mathematica ("Mathematical Principles of Natural Philosophy")*, published in London; also known as Principia, Easton Press, Norwalk Connecticut; printed in 2000.
Nietzsche, F. (1882 and 1887). *The Gay Science*, sections 108 and 125, reproduced in 1974 by Vintage, New York, Walter Kaufmann editor.
Numbers, R.L. (1982). Creationism in 20th-Century America, *Science* (November), pp. 538–544.
Numbers, R.L. (1992). *The Creationists*, Knopf, New York.
Odenwald, S. (2019). *Cosmology*, Arcturus Holdings, London.
Oparin, A.I. (1953). *Origin of Life*, Dover, New York.
Orgel, L.E. (1973). *The Origins of Life*, John Wiley and Sons, New York.
OSBio (2018). *Biology 2e*, 2nd Edition, OpenStax Biology, Rice University, accessed January 2, 2021 from https://d3bxy9euw4e147.cloudfront.net/oscms-prodcms/media/documents/Biology2e-OP_aHSFm3Y.pdf.
OSChem (2019). *Chemistry 2e*, 2nd Edition, OpenStax Chemistry, Rice University, accessed January 16, 2021 from https://d3bxy9euw4e147.cloudfront.net/oscms-prodcms/media/documents/ChemistryAtomsFirst2e-OP_ycRYvTh.pdf.
OSUP (2020). *University Physics*, Volume 3, OpenStax University Physics, Sec. 11.8, LibreStax, accessed December 13, 2020 from https://phys.libretexts.org/Bookshelves/University_Physics/Book%3A_University_Physics_(OpenStax)/Map%3A_University_Physics_III_-_Optics_and_Modern_Physics_(OpenStax)/11%3A_Particle_Physics_and_Cosmology/11.08%3A_Evolution_of_the_Early_Universe.
Padmanabhan, T. (2006). *An Invitation to Astrophysics*, World Scientific, Singapore.
Pais, A. (1982). *Subtle is the Lord …*, Oxford University Press, Oxford, United Kingdom.

Palmer, D. (2009). *Evolution — The Story of Life*, illustrated by P. Barrett, University of California Press, Berkeley, California.

Parrinder, G. (1983). *World Religions*, Facts on File Pub., New York.

Pascal, B. (1979). *Pensées*, Franklin Library, Franklin Center, Pennsylvania.

Pauli, W. (1958). *Theory of Relativity*, Dover, New York.

Pearcey, N.R. and C.B. Thaxton (1994). *The Soul of Science*, Crossway Books, Wheaton, Illinois.

Peebles, P.J.E. (1992). *Quantum Mechanics*, Princeton University Press, Princeton, New Jersey.

Peebles, P.J.E. (1993) *Principles of Physical Cosmology*, Princeton University Press, Princeton, New Jersey.

Pennock, R.T. (1999). *Tower of Babel: The Evidence Against the New Creationism*, MIT Press, Cambridge, Massachusetts.

Perkins, D.H. (1987). *Introduction to High Energy Physics*, 3rd Edition, Addison-Wesley, Reading, Massachusetts.

Perlmutter, S. (2003). Supernovae, Dark Energy, and the Accelerating Universe, *Physics Today* (April), pp. 53–60.

Perlov, D. and A. Vilenkin (2017). *Cosmology for the Curious*, Springer Nature Switzerland, Gewerbestrasse, Switzerland.

Peterson, M., W. Hasker, B. Reichenbach, and D. Basinger (2001). *Philosophy of Religion*, Selected Readings, 2nd Edition, Oxford University Press, New York.

Phillips, R. and S.R. Quake (2006). The Biological Frontier of Physics, *Physics Today* (May), pp. 38–43.

Planck Collaboration (2020). Planck 2018 Results, *Astronomy and Astrophysics*, Volume 641, pp. 1–67, https://doi.org/10.1051/0004-6361/201833910.

Plato (1991). *The Allegory of the Cave, Book VII of The Republic*, Socratic dialogue authored by Plato and published ca. 375 BCE; translation by B. Jowett, Vintage, pp. 253–261; accessed February 4, 2021 from https://www.studiobinder.com/blog/platos-allegory-of-the-cave/.

Ponnamperuma, C. (1974). *Origins*, American Association for the Advancement of Science, Washington, D.C.

Powell, J.L. (2001). *Mysteries of Terra Firma — The Age and Evolution of the Earth*, The Free Press, Simon and Schuster, New York.

Purves, W.K., Sadava, D., Orians, G.H., and Heller, H.C. (2001). *Life — The Science of Biology*, 6th Edition, W.H. Freeman, Sunderland, Massachusetts.

Rauchfuss, H. (2008). *Chemical Evolution and the Origin of Life*, translated by T.N. Mitchell, Springer-Verlag, Berlin.

Raymo, C. (1998). *Skeptics and True Believers*, Walker and Company, New York.

Reese, W.L. (1999). *Dictionary of Philosophy and Religion*, Humanity Books (Prometheus), Amherst, New York.

Rennie, J. (2002). 15 Answers to Creationist Nonsense, *Scientific American* (July), pp. 78–85.

Reynolds, R.J. (2002). The Gas Between the Stars, *Scientific American* (January), pp. 34–43.

Reynolds, S.J., J.K. Johnson, P.J. Morin, and C.M. Carter (2019). *Exploring Geology*, 5th Edition, McGraw-Hill, Boston.

Richter, B. (2006). Theory in Particle Physics: Theological Speculation versus Practical Knowledge, *Physics Today* (October), pp. 8–9.

Ridley, M. (1996). *Evolution*, Blackwell Science, Cambridge, Massachusetts.

Riess, A.G. *et al.* (2016). A 2.4% Determination of the Local Value of the Hubble Constant, *The Astrophysical Journal*, Volume 826, pp. 1–31, July 20, doi:10.3847/0004-637X/826/1/56.

Ross, H. (2001). *The Creator and the Cosmos*, 3rd Edition, NavPress, Colorado Springs, Colorado.

Ross, H. (2016). *Improbable Planet*, Baker Publishing Group, Grand Rapids, Michigan.

Rovelli, C. (2017). *The Order of Time*, Riverhead Books — Penguin Random House, New York.

Rubin, V. (2006). Seeing Dark Matter in the Andromeda Galaxy, *Scientific American* (December), pp. 8–9.

Rubin, V.C., W.K. Ford Jr., N. Thonnard, M.S. Roberts, and J.A. Graham (1976a). Motion of the Galaxy and the Local Group Determined from the Velocity Anisotropy of Distant SC I Galaxies. I. The Data, *Astronomical Journal*, Volume 81 (9, September), pp. 687–718, Bibcode:1976AJ.....81..687R, doi:10.1086/111942.

Rubin, V.C., W.K. Ford Jr., N. Thonnard, M.S. Roberts, and J.A. Graham (1976b). Motion of the Galaxy and the Local Group Determined from the Velocity Anisotropy of Distant SC I Galaxies. II. The Analysis for the Motion, *Astronomical Journal*, Volume 81 (9, September), pp. 719–737, Bibcode:1976AJ.....81..719R, doi:10.1086/111943.

de Rújula, A., H. Georgi, and S.L. Glashow (1975). Hadron Masses in Gauge Theory, *Physical Review D*, Volume 12, pp. 147 ff.

Russell, B. (1972). *A History of Western Philosophy*, Simon and Schuster, New York.
Russell, D.A. (1982). The Mass Extinctions of the Late Mesozoic, *Scientific American* (January), pp. 48–55.
Sagan, C. (1980). *Cosmos*, Random House, New York.
Sambursky, S. (1975). *Physical Thought from the Presocratics to the Quantum Physicists*, Pica Press, New York.
Schroeder, G.L. (1997). *The Science of God*, The Free Press, Simon & Schuster, New York.
Schulze-Makuch, D. and L.N. Irwin (2018). *Life in the Universe*, 3rd Edition, Springer Nature Switzerland, Gewerbestrasse, Switzerland.
Schwarzschild, B. (2003). WMAP Spacecraft Maps the Entire Cosmic Microwave Sky with Unprecedented Precision, *Physics Today* (April), pp. 21–24.
Schwarzschild, B. (2006). New Cosmic Microwave Background Results Strengthen the Case for Inflationary Big Bang Cosmology, *Physics Today* (May), pp. 16–18.
Schwarzschild, B. (2007a). High-Redshift Supernovae Indicate that Dark Energy has been Around for 10 Billion Years, *Physics Today* (January), pp. 21–25.
Schwarzschild, B. (2007b). Three-Dimensional Mapping of Dark Matter Reveals the Expected Filamentary Scaffold, *Physics Today* (March), pp. 20–23.
Scott, E.C. (1997). Antievolution and creationism in the United States, *Annual Review of Anthropology*, Volume 26, pp. 263–289.
SETI-Drake (2020). *Drake Equation*, SETI Institute, accessed January 1, 2021 from https://seti.org/drake-equation-index.
Shapiro, R. (1987). *Origins: A Skeptic's Guide to the Creation of Life on Earth*, Bantam, New York.
Siever, R. (1983). The Dynamic Earth, *Scientific American* (September), pp. 46–55.
Silk, J. (2001). *The Big Bang*, 3rd Edition, Freeman and Company, New York.
Silk, J. (2006). *The Infinite Cosmos*, Oxford University Press, New York.
Singh, S. (2004). *Big Bang*, Fourth Estate, HarperCollins, London.
Skinner, B.J. (1986). Can You Really Believe the Evidence? Two Stories from Geology, *American Scientist* (July–August), pp. 401–409.
Smith, H. (1965). *The Religions of Man*, Harper and Row, New York.

Smith, H. (1991). *The World's Religions*, Harper Collins, San Francisco.

Smolin, L. (2006). *The Trouble with Physics*, Houghton Mifflin, New York.

Smolin, L. (2013). *Time Reborn*, Houghton Mifflin Harcourt, New York.

Smoot, G.F. and D. Scott (2000). Cosmic Background Radiation, *The European Physical Journal C*, Volume 15, pp. 145–149.

Souli, S. (1995). *Greek Mythology*, translated into English by P. Ramp, Toubis, Ilioupoli, Greece.

Sproul, B.C. (1979). *Primal Myths*, Harper Collins, San Francisco.

Srednicki, M. (2000). Dark Matter, *The European Physical Journal C*, Volume 15, pp. 143–144; article written in 1999.

Stahler, S.W. (1991). The Early Life of Stars, *Scientific American* (July), pp. 48–52.

Stallman, J. (2003). John Dewey's New Humanism and Liberal Education for the 21st Century, *Education and Culture*, Fall, Volume 19, Issue 2, pp. 18–22.

Stott, R. (2012). *Darwin's Ghosts — The Secret History of Evolution*, Spiegel & Grau, Random House, New York.

Strange, P. (1998). *Relativistic Quantum Mechanics*, Cambridge University Press, Cambridge, United Kingdom.

Stroll, A. and R.H. Popkins (1972). *Introductory Readings in Philosophy*, Holt, Rinehart and Winston, New York.

Suppes, P. (1957). *Introduction to Logic*, van Nostrand Reinhold, New York.

Surburg, R.F. (1959). *Darwin, Evolution, and Creation*, edited by P.A. Zimmerman, Concordia Publishing House, St. Louis.

Swinburne, R. (2004a). *The Existence of God*, 2nd Edition, Clarendon Press, Oxford University Press, Oxford, United Kingdom.

Swinburne, R. (2004b). *Design Defended, Think: Philosophy for Everyone* (Spring issue), Cambridge University Press, Cambridge, United Kingdom.

Tacitus (ca. 109 CE). *The Annals*, written by Tacitus ca. 109 CE, translated by Alfred John Church and William Jackson Brodribb, printed by Franklin Library in 1982, Franklin Center, Pennsylvania.

Talbot, M. (1981). *Mysticism and the New Physics*, Bantam Books, New York.

Taylor, G.J. (1994). The Scientific Legacy of Apollo, *Scientific American* (July), pp. 40–47.

Taylor, J.R., C.D. Zafiratos, and M.A. Dubson (2004). *Modern Physics for Scientists and Engineers*, 2nd Edition, Prentice Hall, Upper Saddle River, New Jersey.

Tedlock, D. (1996). *Popol Vuh*, Simon & Schuster, New York.

Teerikorpi, P., M. Valtonen, K. Lehto, H. Lehto, G. Byrd, and A. Chernin (2019). *The Evolving Universe and the Origin of Life*, 2nd Edition, Springer Nature Switzerland, Gewerbestrasse, Switzerland.

Thuan, T.X. (1995). *The Secret Melody*, Oxford University Press, Oxford, United Kingdom.

Torah (1992). *Torah*, 3rd Edition, Jewish Publication Society of America, Easton Press, Norwalk, Connecticut.

Tudge, C. (2000a). *The Impact of the Gene*, Hill and Wang, New York.

Tudge, C. (2000b). *The Variety of Life*, Oxford University Press, Oxford, United Kingdom.

Turner, M. (2003). Dark Energy: Just What Theorists Ordered, *Physics Today* (April), pp. 10–11.

Tylor, E.B. (1871). Primitive Culture: Researches into the Development of Mythology, Philosophy, Religion, Art, and Custom, accessed January 26, 2021 from https://archive.org/details/primitiveculture01tylouoft.

Tyson, N. DeGrasse and D. Goldsmith (2004). *Origins*, W.W. Norton, New York.

USGS Continental Drift (2020). Historical Perspective, United States Geological Survey, accessed December 26, 2020 from https://pubs.usgs.gov/gip/dynamic/historical.html.

USGS Time (2007). Divisions of Geologic Time — Major Chronostratigraphic and Geochronologic Units, United States Geological Survey, 2006 Geologic Names Committee, accessed January 7, 2021 from https://pubs.usgs.gov/fs/2007/3015/fs2007-3015.pdf.

van den Bergh, S. and J.E. Hesser (1993). How the Milky Way Formed, *Scientific American* (January), pp. 72–78.

van Over, R. (1980). *Sun Songs: Creation Myths from Around the World*, New American Library, New York.

van Voorst, R.E. (2000). *Jesus Outside the New Testament*, Eerdmans Publishing Company, Grand Rapids, Michigan.

Vidal, G. (1981). *Creation*, Random House, New York.

Wächtershäuser, G. (2006). From Volcanic Origins of Chemoautotrophic Life to Bacteria, Archaea and Eukarya, *Philosophical Transactions of the Royal Society B*, Volume 361, pp. 1787–1808.

Wallace, R.A. (1990). *Biology: The World of Life*, HarperCollins, New York.

Walton, J.H. (2009). *The Lost World of Genesis 1*, InterVarsity Press, Downers Grove, Illinois.

Ward, P.D. and D. Brownlee (2000). *Rare Earth*, Springer-Verlag, New York.

Weinberg, S. (1977). *The First Three Minutes*, 1st Edition, Basic Books, New York.

Weinberg, S. (1992). *Dreams of a Final Theory*, Pantheon, New York.

Weinberg, S. (1993). *The First Three Minutes*, updated edition, Basic Books, New York.

Weinberg, S. (1995). *The Quantum Theory of Fields*, Cambridge University Press, New York, Volume I: Foundations.

Weinberg, S. (1999). *The Quantum Theory of Fields*, Cambridge University Press, New York, Volume III: Supersymmetry.

Weinberg, S. (2008). *Cosmology*, Oxford University Press, Oxford, United Kingdom.

Wells, H.G. (1920). *The Outline of History*, MacMillan Company, New York; released as an ebook by The Gutenberg Project in 2014; accessed February 7, 2021 from http://www.gutenberg.org/files/45368/45368-h/45368-h.htm.

Wheeler, J.A. (2010). *Geons, Black Holes, and Quantum Foam: A Life in Physics*, W. W. Norton & Company, New York, p. 235.

Whorf, B.L. (1987). *Language, Thought, and Reality*, edited by J.B. Carroll, MIT Press, Cambridge, Massachusetts.

Williams, W.S.C. (1991). *Nuclear and Particle Physics*, Oxford University Press, Oxford, United Kingdom.

Witham, L. (2003). *By Design — Science and the Search for God*, Encounter Books, San Francisco.

Woit, P. (2006). *Not Even Wrong*, Perseus Publishing, New York.

Wolff, P. (1965). *Breakthroughs in Physics*, New American Library, New York.

Wong, K. (2002). The Mammals that Conquered the Seas, *Scientific American* (May), pp. 70–79.

Woolfolk, A.E. (2019). *Educational Psychology*, 14th Edition, Pearson Education, New York.

Wu, T-Y. and W-Y P. Hwang (1991). *Relativistic Quantum Mechanics and Quantum Fields*, World Scientific, Singapore.

York, D. (1993). The Earliest History of the Earth, *Scientific American* (January), pp. 90–96.

Zee, A. (2003). *Quantum Field Theory in a Nutshell*, Princeton University Press, Princeton, New Jersey.

Zuckerman, P. (2020), What Is Secular Humanism? A Positive, Ethical, Non-religious Orientation, *Psychology Today*, posted February 12, 2020 and accessed February 7, 2021 from https://www.psychologytoday.com/nz/blog/the-secular-life/202002/what-is-secular-humanism.

Zukav, G. (1979). *The Dancing Wu Li Masters*, William Morrow, HarperCollins, New York.

Zwicky, F. (1933). Die Rotverschiebung von extragalaktischen Nebeln, *Helvetica Physica Acta*, Volume 6, pp. 110–127, Bibcode:1933AcHPh...6..110Z, in German.

Zwicky, F. (1937). On the Masses of Nebulae and of Clusters of Nebulae, *Astrophysical Journal*, Volume 86, pp. 217–246, Bibcode:1937ApJ....86..217Z, doi:10.1086/143864.

Zyla, P.A. *et al.* (2020). Particle Data Group, Review of Particle Physics, *Progress of Theoretical and Experimental Physics*, Volume 2020, Issue 8, August, 083C01, accessed December 7, 2020 from https://doi.org/10.1093/ptep/ptaa104 , see also https://academic.oup.com/ptep/article/2020/8/083C01/5891211.

Selected Abbreviations

Abbreviation	Comment
BCE	Before Common Era
CBR	Cosmic Background Radiation
CDM	Cold, Dark Matter
CMB	Cosmic Microwave Background
COBE	Cosmic Background Explorer
CTI	Cosmic Temple Inauguration
DNA	Deoxyribonucleic Acid
EM	Electromagnetic
GUT	Grand Unified Theory
IARD	International Association for Relativistic Dynamics
IRAS	Infrared Astronomical Satellite
LUCA	Last Universal Common (or Cellular) Ancestor
MACHO	Massive Compact Halo Object
MWI	Many Worlds Interpretation
MYBP	Millions of Years Before Present
NASA	National Aeronautics and Space Administration
NIV	New International Version
NOMA	Nonoverlapping Magisteria
PDG	Particle Data Group
QCD	Quantum Chromodynamics
QED	Quantum Electrodynamics
QFT	Quantum Field Theory
RNA	Ribonucleic Acid

RSV	Revised Standard Version
TOE	Theory of Everything
WIMP	Weakly Interacting Massive Particles
WMAP	Wilkinson Microwave Anisotropy Probe
ΛCDM	Lambda-Cold, Dark Matter

Index

T

Printed in Poland
by Amazon Fulfillment
Poland Sp. z o.o., Wrocław
08 January 2022

6e9859bc-4bc0-400e-a2b5-e9852486bb12R01